General Theory of Relativity From Basics

Sasan Ardalan, Ph.D.

https://www.scientificworks.org

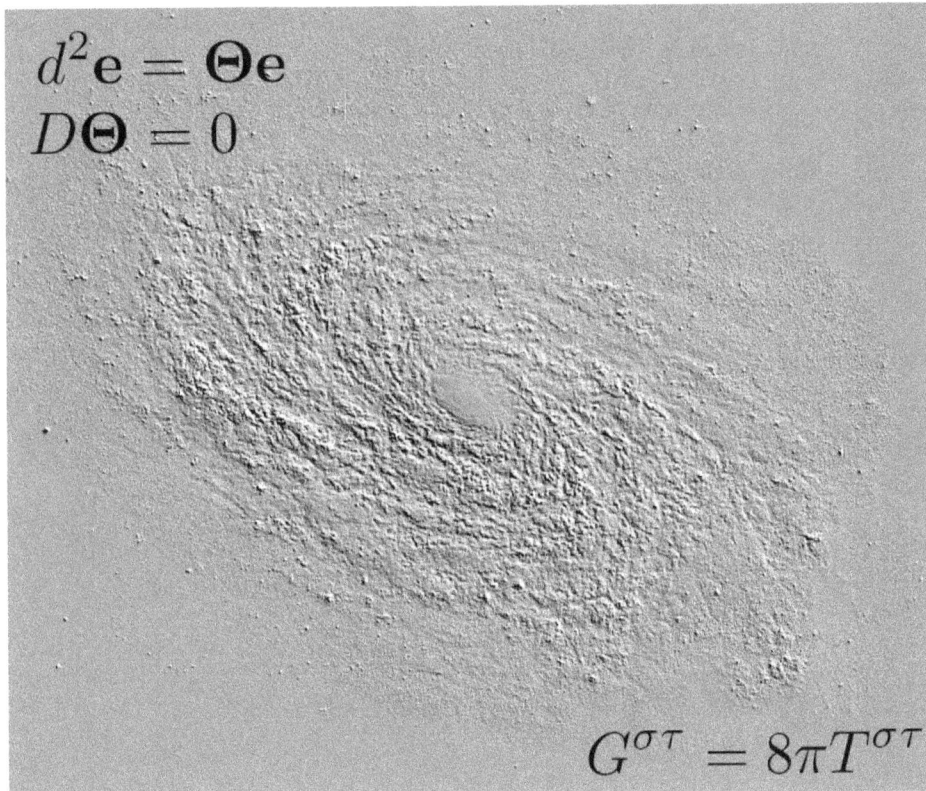

$$d^2\mathbf{e} = \mathbf{\Theta}\mathbf{e}$$
$$D\mathbf{\Theta} = 0$$

$$G^{\sigma\tau} = 8\pi T^{\sigma\tau}$$

Dusty Spiral Galaxy NGC 4414
Image Credit: The Hubble Heritage Team (AURA/STScI/NASA)
http://hubblesite.org/gallery/album/galaxy_collection/pr1999025a/

Sasan Ardalan received his Diploma from Alborz, Tehran, Iran. He received his Ph.D. from North Carolina State University in 1983 in Electrical Engineering. He was an Associate Professor at NC State in 1991 and an Adjunct Professor at Duke University. He has published many articles in refereed journals in electronics. He has 12 issued US patents. He holds an Extra Class Amateur Radio license AJ7BF.

Contents

List of Figures

Chapter 1

Introduction

The purpose of this book is to provide the derivation of Einstein's Geometric Theory of Gravity starting from the basics. Through tensor analysis and differential forms we will derive the Bianchi identity. We then tie the conservation of energy-momentum to the "automatically" conserved feature of the geometry. We will derive:

$$G^{\sigma\tau} = 8\pi T^{\sigma\tau} \tag{1}$$

As with the referenced books, this book must be read and re-read to make sense. But in absolutely no case do we skip a step so that you have all the material needed without wondering what mountain was skipped and where did this come from. Again its all here with references. In only some cases do we show an equation without a derivation but we point out that the full derivation appears in a later chapter in a more elegant form.

Chapter 2

Tensors in Euclidian Space

We are going to look into invariants under coordinate transformation and also curvilinear coordinate systems. This material will lead to Tensors.

We start by examining a vector **A** that is tangent to a parametric curve in three dimensions at point P. See Figure 1 . Let $\mathbf{e_1}, \mathbf{e_2}, \mathbf{e_3}$ denote the unit vectors in the x, y, z direction in E^3. These are the same as the familiar unit vectors, $\mathbf{i}, \mathbf{j}, \mathbf{k}$. Let the projection of the vector $\mathbf{A}$ on the unit vectors be A_1, A_2, A_3. Then the length of $\mathbf{A}$ is:

$$|\mathbf{A}|^2 = A_1^2 + A_2^2 + A_3^2 \tag{2}$$

Now we want to define a new set of orthonormal coordinate vectors but associated with curvilinear coordinates. We want to express the length of $\mathbf{A}$ in this coordinate system. The length should be invariant under the new coordinate system. Consider spherical polar coordinates. In Figure 2, the coordinates of point P in rectangular coordinates are:

$$
\begin{aligned}
x^1 &= \rho \sin \theta \cos \phi \\
x^2 &= \rho \sin \theta \sin \phi \\
x^3 &= \rho \cos \theta
\end{aligned}
\tag{3}
$$

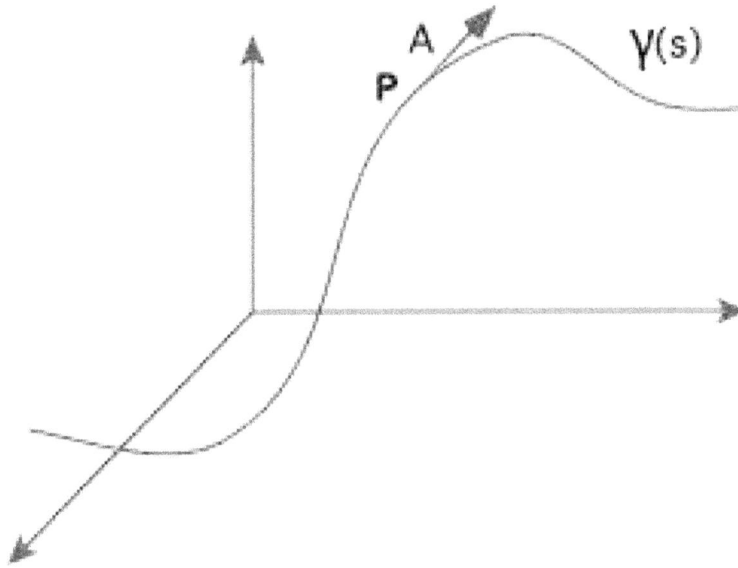

Figure 1: Parametric Curve with Tangent Vector A at point P

where the notation $x^i, i = 1, 2, 3$ denotes the rectangular coordinates and is a prelude to notation to be introduced with tensors.

At point P we construct the three orthonormal vectors $\epsilon_1, \epsilon_2, \epsilon_3$ tangent to the coordinate curves $\rho = c_1, \theta = c_2, \phi = c_3$. The components are given by,

$$
\begin{aligned}
\epsilon_1 &= (\sin \theta \cos \phi, \sin \theta \sin \phi, \cos \theta) \\
\epsilon_2 &= (\cos \theta \cos \phi, \cos \theta \sin \phi, -\sin \theta) \\
\epsilon_3 &= (-\sin \phi, \cos \phi, 0)
\end{aligned}
\tag{4}
$$

6

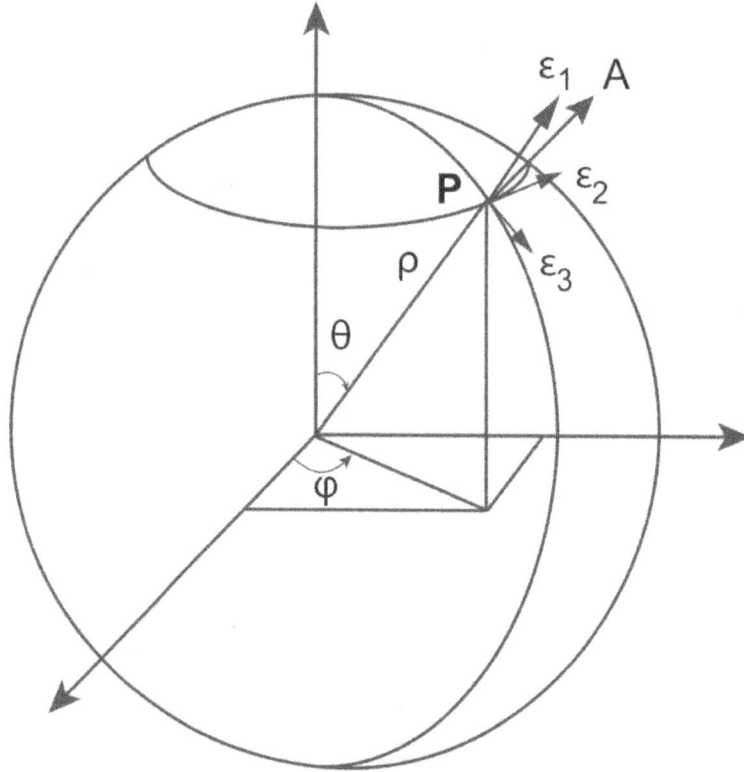

Figure 2: Spherical Coordinates and Orthonormal Basis Coordinates at P

Now consider an arbitrary displacement $d\mathbf{r}$ from the position P. See Figure 3. In rectangular coordinates $d\mathbf{r} = dx^1\mathbf{e_1} + dx^2\mathbf{e_2} + dx^3\mathbf{e_3}$. How do we express $d\mathbf{r}$ in terms of $d\rho, d\theta, d\phi$? We need to project $d\mathbf{r}$ unto $\epsilon_1, \epsilon_2, \epsilon_3$. We get,

$$d\mathbf{r} = d\rho\epsilon_1 + \rho d\theta\epsilon_2 + \rho\sin\theta d\phi\epsilon_3 \tag{5}$$

Let the vector $\mathbf{A}$ be the tangent to the parametric curve $\gamma(s)$ with parameter s. Then in rectangular coordinates:

$$\mathbf{A} = \frac{d\mathbf{r}}{ds} = \frac{dx^1}{ds}\mathbf{e_1} + \frac{dx^2}{ds}\mathbf{e_2} + \frac{dx^3}{ds}\mathbf{e_3} \tag{6}$$

Define $A_1 = \frac{dx^1}{ds}, A_2 = \frac{dx^1}{ds}, A_3 = \frac{dx^1}{ds}$. Also define $\bar{A}_1 = \frac{d\rho}{ds}, \bar{A}_2 = \frac{d\theta}{ds}, \bar{A}_3 = \frac{d\phi}{ds}$. Then,

$$\mathbf{A} = \bar{A}_1\epsilon_1 + \rho\bar{A}_2\epsilon_2 + \rho\sin\theta\bar{A}_3\epsilon_3 \tag{7}$$

It follows from (7) that,

7

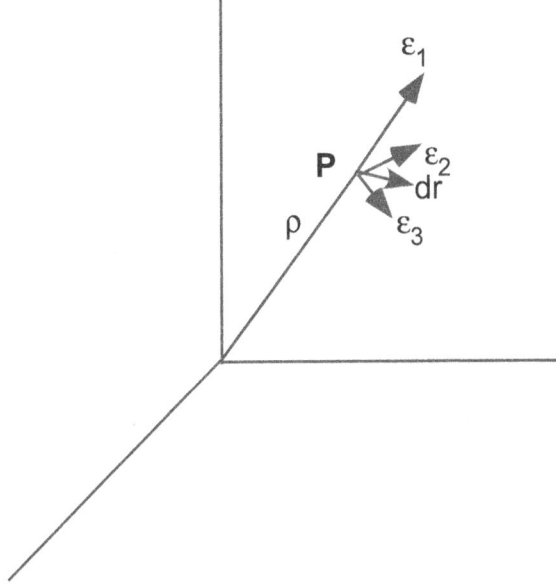

Figure 3: Differential $d\mathbf{r}$ from Point P

$$\bar{A}_1 = \mathbf{A} \cdot \epsilon_1$$
$$\rho \bar{A}_2 = \mathbf{A} \cdot \epsilon_2$$
$$\rho \sin \theta \bar{A}_3 = \mathbf{A} \cdot \epsilon_3 \tag{8}$$

Based on (4),

$$
\begin{aligned}
\bar{A}_1 &= \sin\theta\cos\phi A_1 + \sin\theta\sin\phi A_2 + \cos\theta A_3 \\
\bar{A}_2 &= \frac{1}{\rho}\left(\cos\theta\cos\phi A_1 + \cos\theta\sin\phi A_2 - \sin\theta\right) \\
\bar{A}_3 &= -\frac{sin\phi}{\rho\sin\theta}A_1 + \frac{\cos\phi}{\rho\sin\theta}A_2
\end{aligned}
\tag{9}
$$

Since $\epsilon_1, \epsilon_2, \epsilon_3$ are orthonormal, then the length of $\mathbf{A}$ is(see(7)):

$$|\mathbf{A}|^2 = \bar{A}_1^2 + \rho^2 \bar{A}_2^2 + \rho^2 \sin^2\theta \bar{A}_3^2 \tag{10}$$

The length is invariant so the length computed by (2) and (10) are the same.

Define the Matrix,

$$\bar{G} = diag(1, \rho^2, \rho^2 \sin^2 \theta) \tag{11}$$

Then,

$$|\mathbf{A}|^2 = [\bar{A}_1 \ \bar{A}_2 \ \bar{A}_3]\bar{G}[\bar{A}_1 \ \bar{A}_2 \ \bar{A}_3]^T \tag{12}$$

For rectangular coordinates, define $G = diag(1, 1, 1)$. Then,

$$|\mathbf{A}|^2 = [A_1 \ A_2 \ A_3]G[A_1 \ A_2 \ A_3]^T \tag{13}$$

Also in terms of $d\mathbf{r}$, we have,

$$|d\mathbf{r}|^2 = [dx^1 \ dx^2 \ dx^3]G[dx^1 \ dx^2 \ dx^3]^T \tag{14}$$

and

$$|d\mathbf{r}|^2 = [d\rho \ d\theta \ d\phi]\bar{G}[d\rho \ d\theta \ d\phi]^T \tag{15}$$

We will now derive the above via another formulation.

Define,

$$\begin{aligned} \bar{x}^1 &= \rho \\ \bar{x}^2 &= \theta \\ \bar{x}^3 &= \phi \end{aligned} \tag{16}$$

Now, the spherical coordinates are related to the rectangular coordinate:

$$\begin{aligned} \bar{x}^1 &= x^1(x^i) \\ \bar{x}^2 &= x^2(x^i) \\ \bar{x}^3 &= x^3(x^i) \end{aligned} \tag{17}$$

This notation is to be interpreted as follows: $\bar{x}^1$ which is ρ, is a function of x^1, x^2, x^3 the rectangular coordinates (x, y, z) through a potentially nonlinear function $x^1(x^1, x^2, x^3)$ which we abbreviate $x^1(x^i), i = 1, 2, 3$. We could have used a function $f^1(x^1, x^2, x^3)$, but the notation does not create ambiguity.

Now, (8) suggest that $\bar{A}_i$ are related to A_i via a 3×3 matrix. That matrix relationship is:

$$\bar{A}_j = \sum_{h=1}^{n} \frac{\partial \bar{x}^j}{\partial x^h} A_h \tag{18}$$

Now refering to (3) we have,

9

$$\begin{aligned}
x^1 &= \bar{x}^1 \sin \bar{x}^2 \cos \bar{x}^3 \\
x^2 &= \bar{x}^1 \sin \bar{x}^2 \sin \bar{x}^3 \\
x^3 &= \bar{x}^1 \cos \bar{x}^2
\end{aligned} \tag{19}$$

Following [8], the matrix of derivatives $\partial x^h / \partial \bar{x}^j$ of this transformation is,

$$\left(\frac{\partial x^h}{\partial \bar{x}^j} \right) = \begin{pmatrix} \sin \bar{x}^2 \cos \bar{x}^3 & \bar{x}^1 \cos \bar{x}^2 \cos \bar{x}^3 & -\bar{x}^1 \sin \bar{x}^2 \sin \bar{x}^3 \\ \sin \bar{x}^2 \sin \bar{x}^3 & \bar{x}^1 \cos \bar{x}^2 \sin \bar{x}^3 & \bar{x}^1 \sin \bar{x}^2 \cos \bar{x}^3 \\ \cos \bar{x}^2 & -\bar{x}^1 \sin \bar{x}^2 & 0 \end{pmatrix} \tag{20}$$

Also,

$$\left(\frac{\partial \bar{x}^j}{\partial x^h} \right) = \begin{pmatrix} \sin \bar{x}^2 \cos \bar{x}^3 & \sin \bar{x}^2 \sin \bar{x}^3 & \cos \bar{x}^2 \\ \frac{\cos \bar{x}^2 \cos \bar{x}^3}{\bar{x}^1} & \frac{\cos \bar{x}^2 \sin \bar{x}^3}{\bar{x}^1} & -\frac{\sin \bar{x}^2}{\bar{x}^1} \\ -\frac{\sin \bar{x}^3}{\bar{x}^1 \sin \bar{x}^2} & \frac{\cos \bar{x}^3}{\bar{x}^1 \sin \bar{x}^2} & 0 \end{pmatrix} \tag{21}$$

"Thus, substituting from (21) in (18) we infer that, according to this prescription, the components $\bar{A}_j$ of the vector $\mathbf{A}$ are given in spherical polar coordinates by,"[8]

$$\begin{aligned}
\bar{A}_1 &= \sin \bar{x}^2 \cos \bar{x}^3 A_1 + \sin \bar{x}^2 \sin \bar{x}^3 A_2 + \cos \bar{x}^2 A_3 \\
\bar{A}_2 &= \frac{\cos \bar{x}^2 \cos \bar{x}^3}{\bar{x}^1} A_1 + \frac{\cos \bar{x}^2 \sin \bar{x}^3}{\bar{x}^1} A_2 - \frac{\sin \bar{x}^2}{\bar{x}^1} A_3 \\
\bar{A}_3 &= -\frac{\sin \bar{x}^3}{\bar{x}^1 \sin \bar{x}^2} A_1 + \frac{\cos \bar{x}^3}{\bar{x}^1 \sin \bar{x}^2} A_2
\end{aligned} \tag{22}$$

Consider the scalar function $\phi(x^h)$ of the coordinates x^h. This function is said to be a scalar or invariant under the transformation (17) if its transform $\bar{\phi}(\bar{x}^j)$ possesses the same numerical value:

$$\bar{\phi}(\bar{x}^j) = \phi(x^h) \tag{23}$$

Both sets of coordinates refer to the same point P of E^3.

The gradient of the scalar function $\phi(x^h)$ is defined by the vector $\partial \phi / \partial x^h$. For reference, let $\phi(x, y, z)$ be a scalar function in rectangular coordinates. Then $\nabla \phi(x, y, z) = \frac{\partial \phi}{\partial x} \mathbf{i} + \frac{\partial \phi}{\partial y} \mathbf{j} + \frac{\partial \phi}{\partial z} \mathbf{k}$.

Now lets differentiate $\bar{\phi}(\bar{x}^j)$ with respect to $\bar{x}^j$ to obtain the gradient. Using the chain rule:

$$\frac{\partial \bar{\phi}}{\partial \bar{x}^j} = \sum_{h=1}^{n} \frac{\partial x^h}{\partial \bar{x}^j} \frac{\partial \phi}{\partial x^h} \tag{24}$$

As a tensor:

$$\frac{\partial \bar{\phi}}{\partial \bar{x}^j} = \frac{\partial x^h}{\partial \bar{x}^j} \frac{\partial \phi}{\partial x^h} \tag{25}$$

10

Note the dummy index h.

It would be useful to write

$$\nabla \bar{\phi} = \frac{\partial x^h}{\partial \bar{x}^j} \nabla \phi \tag{26}$$

Now lets consider the tangent to the curve at point P. The curve is defined by $x^i = x^i(s)$. The tangent vector field $\mathbf{T} = (T^i)$ is defined by,

$$T^i = \frac{dx^i}{ds} \tag{27}$$

To express the same curve in the barred coordinate system,

$$\bar{x}^i = \bar{x}^i(t). \tag{28}$$

The tangent vector is:

$$\bar{T}^i = \frac{d\bar{x}^i}{ds} \tag{29}$$

By the chain rule we can write,

$$\frac{d\bar{x}^i}{ds} = \frac{\partial \bar{x}^i}{\partial x^r} \frac{dx^r}{ds} \tag{30}$$

Or,

$$\bar{T}^i = T^r \frac{\partial \bar{x}^i}{\partial x^r} \tag{31}$$

Operation	Type	Relation
Tangent	Contravariant	$\frac{d\bar{x}^i}{ds} = \frac{\partial \bar{x}^i}{\partial x^r} \frac{dx^r}{ds}$
Gradient	Covariant	$\nabla \bar{\phi} = \frac{\partial x^h}{\partial \bar{x}^j} \nabla \phi$

In general:

Type	Relation
Contravariant	$\bar{T}^i = T^r \frac{\partial \bar{x}^i}{\partial x^r}$
Covariant	$\bar{T}_i = T_j \frac{\partial x^j}{\partial \bar{x}^i}$

Note that for contravariant tensor the index is placed as superscript and for covariant it is placed as subscript.

A very important property of contravariant and covariant tensors is that their inner product is an invariant.

Thus,

$$\bar{T}^j \bar{T}_l = \frac{\partial \bar{x}^j}{\partial x^h} \frac{\partial x^k}{\partial \bar{x}^l} T^h T_k \tag{32}$$

11

Lets perform the process of contraction on j and l where we let $j = l$, i.e., we form an inner product,

$$\bar{T}^j \bar{T}_j = \delta_{hk} T^h T_k = T^k T_k \tag{33}$$

where we have used

$$\sum_{j=1}^{n} \frac{\partial \bar{x}^j}{\partial x^h} \frac{\partial x^k}{\partial \bar{x}^j} = \delta_k^h \tag{34}$$

Note that for δ_k^h the inner index k represents row and the outer index h represents column.

As[8] points out this result is indeed very satisfactory, as an invariant resembling an inner product has been constructed.

A vector is represented as :

$$\bar{A}_i = [\bar{A}_1 \ \bar{A}_2 \ \bar{A}_3]^T \tag{35}$$

Now with $\bar{G}$ the matrix defined in (11), we define,

$$\bar{A}^i = \bar{G}[\bar{A}_1 \ \bar{A}_2 \ \bar{A}_3]^T \tag{36}$$

Or,

$$\bar{A}^i = \bar{g}_{ij} \bar{A}_j \tag{37}$$

Then the length squared of the vector $\bar{A}$ can be expressed as,

$$|\bar{A}|^2 = \bar{A}_i \bar{A}^i \tag{38}$$

Of course the length is invariant, $|\bar{A}|^2 = |A|^2$ and,

$$|A|^2 = A_i A^i \tag{39}$$

So again the inner product of a contravariant and covariant vector yield an invariant.

Now, referring to (37) we see that the matrix $\bar{g}_{ij}$ can be used to raise the index of a vector. The matrix $\bar{g}_{ij}$ is referred to as the metric.

Also the inverse of the metric is denoted as,

$$\bar{g}^{ij} = \bar{G}^{-1} \tag{40}$$

And,

$$g_{ij} \bar{g}^{ij} = \delta_{ij} \tag{41}$$

We can write the arc lengths as,

$$|d\mathbf{r}|^2 = \bar{g}_{ij} d\bar{x}^i d\bar{x}^j \tag{42}$$

12

and,

$$|d\mathbf{r}|^2 = g_{ij}dx^i dx^j \tag{43}$$

Covariant Differentiation

Consider the partial differentiation of the transformation law,

$$\bar{T}_i = T_r \frac{\partial x^r}{\partial \bar{x}^i} \tag{44}$$

Thus,

$$\frac{\partial \bar{T}_i}{\partial \bar{x}^k} = \frac{\partial T_j}{\partial \bar{x}^k}\frac{\partial x^j}{\partial \bar{x}^i} + T_r \frac{\partial^2 x^r}{\partial \bar{x}^k \partial \bar{x}^i} \tag{45}$$

Now if the term $\frac{\partial^2 x^r}{\partial \bar{x}^k \partial \bar{x}^i} = 0$ then, $\frac{\partial \bar{T}_i}{\partial \bar{x}^k} = \frac{\partial T_j}{\partial \bar{x}^k}\frac{\partial x^j}{\partial \bar{x}^i}$ transforms as tensor. However, this only holds true in, for example, rectangular coordinate systems. It does not hold true in curvilinear coordinate systems. So we have a problem. To resolve this problem we need an expression for $\frac{\partial^2 x^r}{\partial \bar{x}^k \partial \bar{x}^i}$ which should be related to the metric for the curvalinear coordinate system.

Christoffel Symbols

In order to proceed we define the Christoffel symbols of the first kind:

$$\Gamma_{ijk} = \frac{1}{2}\left[\frac{\partial}{\partial x^i}(g_{jk}) + \frac{\partial}{\partial x^j}(g_{ki}) - \frac{\partial}{\partial x^k}(g_{ij})\right] \tag{46}$$

The symbols are related to the metric and their significance will become apparent in the following developments. Although at this point we define the Christoffel symbols as indicated we will show how they arise in the development of differential forms in later chapters. For now we are trying to resolve the problem with differentiation outlined above.

A good example for the computation of the Christoffel symbols is provided in [6] from the metric of spherical coordinates. In this case,

$$G = \begin{bmatrix} 1 & 0 & 0 \\ 0 & (x^1)^2 & 0 \\ 0 & 0 & (x^1)^2 \sin^2 x^2 \end{bmatrix} \tag{47}$$

$$
\begin{aligned}
\Gamma_{221} &= -x^1 \\
\Gamma_{212} &= -x^1 \\
\Gamma_{122} &= -x^1 \\
\Gamma_{331} &= -x^1 \sin^2 x^2 \\
\Gamma_{323} &= (x^1)^2 \sin x^2 \cos x^2 \\
\Gamma_{332} &= -(x^1)^2 \sin x^2 \cos x^2 \\
\Gamma_{133} &= x^1 \sin^2 x^2 \\
\Gamma_{313} &= x^1 \sin^2 x^2 \\
\Gamma_{233} &= (x^1)^2 \sin x^2 \cos x^2
\end{aligned}
\tag{48}
$$

Note all other terms are zero.

Basic properties of the Christoffel symbols of the first kind are [6]:

(i) $\Gamma_{ijk} = \Gamma_{jik}$ (Symmetry in the first two indices)

(ii) all Γ_{ijk} vanish if all g_{ij} are constant

A usful formula:
$$
\frac{\partial g_{ik}}{\partial x^j} = \Gamma_{ijk} + \Gamma_{jki}
\tag{49}
$$

Thus in any coordinate system in which the metric tensor has constant components the Christoffel symbols uniformly vanish.

Transformation Law   See [6] for derivation of the following:

$$
\bar{\Gamma}_{ijk} = \Gamma_{rst} \frac{\partial x^r}{\partial \bar{x}^i} \frac{\partial x^s}{\partial \bar{x}^j} \frac{\partial x^t}{\partial \bar{x}^k} + g_{rs} \frac{\partial^2 x^r}{\partial \bar{x}^i \partial \bar{x}^j} \frac{\partial x^s}{\partial \bar{x}^k}
\tag{50}
$$

Christoffel Symbols of the Second Kind

Definition:

$$
\Gamma^i_{jk} = y^{ir} \Gamma_{jkr}
\tag{51}
$$

Similar to raising the third subscript but we are not dealing with a tensor. Basic properties carry over from Γ_{ijk}:

(i) $\Gamma^i_{jk} = \Gamma^i_{kj}$ (Symmetry in the first two indices)

(ii) all Γ^i_{jk} vanish if all g_{ij} are constant

14

Transformation Law See [6] for derivation of the following:

$$\bar{\Gamma}^i_{jk} = \Gamma^r_{st} \frac{\partial x^i}{\partial \bar{x}^r} \frac{\partial x^s}{\partial \bar{x}^j} \frac{\partial x^t}{\partial \bar{x}^k} + \frac{\partial^2 x^r}{\partial \bar{x}^j \partial \bar{x}^k} \frac{\partial x^i}{\partial \bar{x}^r} \tag{52}$$

An Important Formula

$$\frac{\partial^2 x^r}{\partial \bar{x}^i \partial \bar{x}^j} = \bar{\Gamma}^s_{ij} \frac{\partial x^r}{\partial \bar{x}^s} - \Gamma^r_{st} \frac{\partial x^s}{\partial \bar{x}^i} \frac{\partial x^t}{\partial \bar{x}^j} \tag{53}$$

We are now in a position to address Covariant Differentiation. In (45) substitute for $\frac{\partial^2 x^r}{\partial \bar{x}^i \partial \bar{x}^j}$ based on (53):

$$\begin{aligned}
\frac{\partial \bar{T}_i}{\partial \bar{x}^k} &= \frac{\partial T_j}{\partial \bar{x}^k} \frac{\partial x^r}{\partial \bar{x}^i} \frac{\partial x^s}{\partial \bar{x}^k} + T_r \left(\bar{\Gamma}^s_{ij} \frac{\partial x^r}{\partial \bar{x}^s} - \Gamma^r_{st} \frac{\partial x^s}{\partial \bar{x}^i} \frac{\partial x^t}{\partial \bar{x}^k} \right) \\
&= \frac{\partial T_r}{\partial \bar{x}^s} \frac{\partial x^r}{\partial \bar{x}^i} \frac{\partial x^s}{\partial \bar{x}^k} + \bar{\Gamma}^t_{ik} \bar{T}_t - \Gamma^t_{rs} T_t \frac{\partial x^r}{\partial \bar{x}^i} \frac{\partial x^s}{\partial \bar{x}^k}
\end{aligned} \tag{54}$$

which rearranges to

$$\frac{\partial \bar{T}_i}{\partial \bar{x}^k} - \bar{\Gamma}^t_{ik} \bar{T}_t = \left(\frac{\partial T_r}{\partial \bar{x}^s} - \Gamma^t_{rs} T_t \right) \frac{\partial x^r}{\partial \bar{x}^i} \frac{\partial x^s}{\partial \bar{x}^k} \tag{55}$$

This is a defining law of a covariant tensor of order two. Thus by subtracting linear combinations of the components of **T** from itself we get a tensor. These corrections are related to the metric via the Christoffel symbols.

Definition

The Covariant Derivative with respect to x^k of a covariant vector $\mathbf{T_i}$ is the tensor

$$\nabla_k T_i = \left(\frac{\partial T_i}{\partial x^k} - \Gamma^t_{ik} T_t \right) \tag{56}$$

Definition

The Contravariant Derivative with respect to x^k of a contravariant vector $\mathbf{T^k}$ is the tensor

$$\nabla_k T^i = \left(\frac{\partial T^i}{\partial x^k} + \Gamma^i_{tk} T^t \right) \tag{57}$$

From [13] :

15

Transformations and Tensors

The major objective of tensor analysis is to determine algabraic representations for physical or geometric relations in a form independent of coordinate system, that is, we look for the algabraic and geometric invariants of a given transformation group.

In addition to this comment, [13] adds,"in considering the processes of differentiation in a Riemannian space, the concept of invariance should be kept in the foreground. The general form of a geometric or physiscal law is independent of coordinate system when expressed entirely in terms of tensors."

From [13] :

Riemannian Geometry

Definition
An n-space endowed with a covariant tensor g_{jk} of second order, which is symmetric, is said to be a Riemannian space. The geometry of the space is said to be Riemannian.

Definition A quadratic differential form

$$ds^2 = g_{jk}dx^j dx^k \tag{58}$$

can be asociated with the tensor g_{jk}. It is called the fundamental metric form of the space and g_{jk} is said to be the fundamental metric tensor.

Note that we do not require that ds^2 be positive definite in anticipation of General Relativity.

Conjugate or Associated Metric Tensor For the metric g_{jk} the tensor g^{jk} for which,

$$g_{jk}g^{kp} = \delta_j{}^p \tag{59}$$

$$g_{kj}g^{pk} = \delta_j{}^p \tag{60}$$

is defined as the conjugate or associated metric tensor.

Raising and Lower Indices If T^{ij} are the components of a given tensor, then

$$T_k^j = g_{ki}T^{ij} \tag{61}$$

$$T_{kp} = g_{ki}g_{pj}T^{ij} \tag{62}$$

Note that although the tensors $\{T^{ij}\}\{T_{ij}\}, \{T_i^j\}, \{T^i{}_j\}$ are distinct from the algebraic point of view, each represents the same geometric or physical entity [13].

16

We introduce the notation $\partial_\mu = \frac{\partial}{\partial x^\mu}$.

With the covariant divergence of V^μ given by,

$$\nabla_\mu V^\mu = \partial_\mu V^\mu + \Gamma^\mu_{\mu\lambda} V^\lambda \tag{63}$$

It can be shown that ,

$$\Gamma^\mu_{\mu\lambda} = \frac{1}{\sqrt{|g|}} \partial_\lambda \sqrt{|g|} \tag{64}$$

where $|g|$ is the determinant of g_{ij}.
Thus,

$$\nabla_\mu V^\mu = \frac{1}{\sqrt{|g|}} \partial_\mu (\sqrt{|g|} V^\mu). \tag{65}$$

Some Applications

Laplacian

Consider a scalar function ϕ. Its gradient is a covariant vector $\nabla_i \phi$. The contravariant form can be obtained by raising the index, $g^{ij} \nabla_i \phi$. In Cartesian coordinates both the covariant and contravariant vector are equal to $\nabla \phi$. Take the contravariant derivative of $g^{ij} \nabla_i \phi$ to obtain:

$$\nabla_i g^{ij} \nabla_j \phi \tag{66}$$

Now take the trace,

$$\nabla^2 \phi = \frac{\partial}{\partial x^j} \left(g^{ij} \frac{\partial \phi}{\partial x^i} \right) + \Gamma^j_{jk} g^{ik} \frac{\partial \phi}{\partial x^i} \tag{67}$$

But,

$$\Gamma^j_{jk} g^{ik} = \frac{1}{\sqrt{|g|}} \frac{\partial \sqrt{|g|}}{\partial x^j} g^{ik} \tag{68}$$

Leading to

$$\nabla^2 \phi = \frac{1}{\sqrt{|g|}} \frac{\partial}{\partial x^j} \left\{ \sqrt{|g|} g^{ij} \frac{\partial \phi}{\partial x^i} \right\} \tag{69}$$

In orthogonal coordinates the metric is diagonal ($g_{ii} = h_i, g_{ik} = 0, i \neq j$). Hence,

$$\nabla^2 \phi = \frac{1}{h_1 h_2 h_3} \left[\frac{\partial}{\partial x^1} \left(\frac{h_2 h_3}{h_1} \frac{\partial \phi}{\partial x^1} \right) + \frac{\partial}{\partial x^2} \left(\frac{h_3 h_1}{h_2} \frac{\partial \phi}{\partial x^2} \right) + \frac{\partial}{\partial x^3} \left(\frac{h_1 h_2}{h_3} \frac{\partial \phi}{\partial x^3} \right) \right] \tag{70}$$

For spherical coordinates,

$$G = \begin{bmatrix} 1 & 0 & 0 \\ 0 & (x^1)^2 & 0 \\ 0 & 0 & (x^1)^2 \sin^2 x^2 \end{bmatrix} \tag{71}$$

17

In terms of (r, θ, ϕ), $h_1 = 1, h_2 = r^2, h_3 = r^2 \sin \theta, \sqrt{g} = r^2 \sin \theta$ the Laplacian is,

$$\nabla^2 f = \frac{1}{r^2} \frac{\partial}{\partial r} \left(r^2 \frac{\partial f}{\partial r} \right) + \frac{1}{r^2 \sin \theta} \frac{\partial}{\partial \theta} \left(\sin \theta \frac{\partial f}{\partial \theta} \right) + \frac{1}{r^2 \sin^2 \theta} \frac{\partial^2 f}{\partial \phi^2} \tag{72}$$

Divergence

The divergence is,

$$\nabla_\mu V^\mu = \frac{1}{\sqrt{|g|}} \partial_\mu (\sqrt{|g|} V^\mu). \tag{73}$$

For (x^1, x^2, x^3) we use (r, θ, ϕ) as before. We have $\sqrt{|g|} = r^2 \sin \theta$. Now the components (V_1, V_2, V_3) correspond to $V_r, \frac{V_\theta}{r}, \frac{V_\phi}{r \sin \theta}$. See (7). Then,

$$\nabla_\mu V^\mu = \frac{1}{r^2 \sin \theta} \left(\sin \theta \frac{\partial (r^2 V_r)}{\partial r} + r \frac{\partial (V_\theta \sin \theta)}{\partial \theta} + r \frac{\partial V_\phi}{\partial \phi} \right) \tag{74}$$

Finally,

$$\nabla_\mu V^\mu = \frac{1}{r^2} \frac{\partial (r^2 V_r)}{\partial r} + \frac{1}{r \sin \theta} \frac{\partial (V_\theta \sin \theta)}{\partial \theta} + \frac{1}{r \sin \theta} \frac{\partial V_\phi}{\partial \phi} \tag{75}$$

Chapter 3

Parallel Transport

Consider the vector U at a on the curve in the two dimensional space shown in Figure 4. Now using the construction shown parallel transport U to the position b. Let the interval $a - b$ be very small. Note that the length of U does not change. Now consider Figure 5. In this case we have transported vector U along the closed curve, and, as shown it ends up right on top of itself at a. Note also that the path of the tip of U also depends on the curve. Now consider if we started out at f with vector U. Now if we parallel transported U along the path $f - e - d - c - b - a$ to a or if we took the path along $f - g - h - a$ in this case the vector will coincide at a. So in flat space it turns out that the parallel transport of a vector along two curves to a common point will end up with the vectors unchanged at that point independent of the path taken. This is a property of flat space.

By the way, we are using the "Shild's Ladder" construction, see [3] Box 10.2, over small incremental (differential) points along the curves to parallel transport the vector.

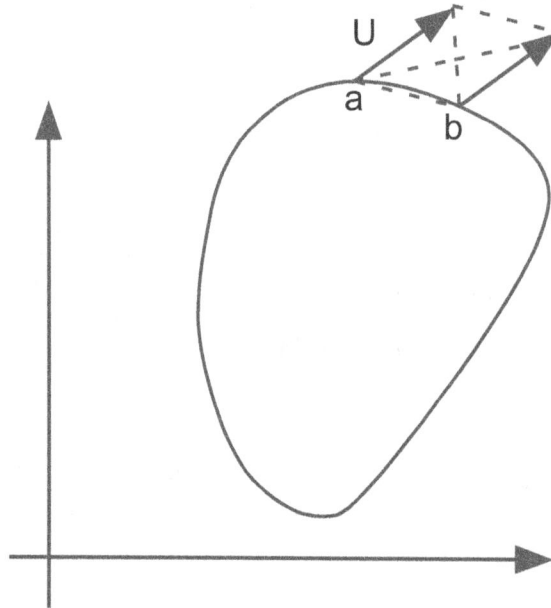

Figure 4: Parallel Transport

Now consider the curved space show in Figure 6. Obviously we are dealing with more than two dimensions in this case. Now the vector at P_1 is $X^j_{(1)}$. We have two paths to the point P_2. One is along curve C and the other is along curve C^*. Using the "Shild's Ladder" construction along path C parallel transport $X^j_{(1)}$ from P_1 to P_2. Now do the same along path C^*. The vector $X^j_{(1)}$ ends up with a different orientation at P_2 than the same vector parallel transported along path C. This is a property of curved space (later curved spacetime). It is important to note that the final orientation of the vector parallel transported from point P_1 to point P_2 depends on the curve along which it is transported. This is very important. Since, in the development of the covariant derivative to be presented next, it will be clear that the derivative is taken with respect to a curve (or path) in the Euclidean space.

Figure 6 is adapted from [8].

Covariant Derivative

Consider the n-dimensional space shown in Figure 7. Also shown is a curve defined parametrically at the point $P(\lambda)$ with parameter λ. The vector $\mathbf{v}$ is defined at λ_0 on the curve. Now, obtain the value of the vector field $\mathbf{v}$ at the

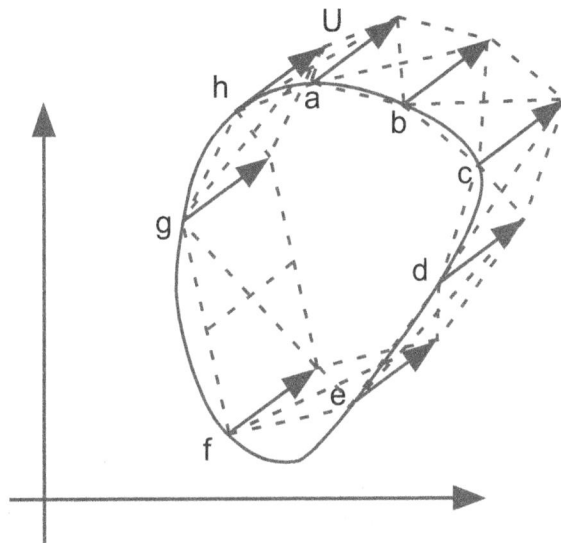

Figure 5: Parallel Transport in Flat Space

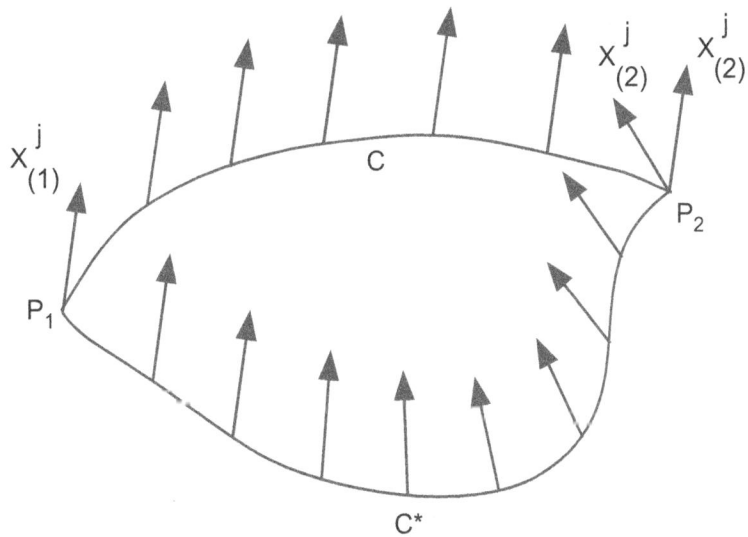

Figure 6: Parallel Transport Curved Space

point $P(\lambda+\epsilon)$. This is $\mathbf{v}(\lambda_0+\epsilon)$. Now parallel transport the vector back to the point $P(\lambda_0)$. See Figure 7. The change in the vector is $\nabla_{\mathbf{u}}\mathbf{v}$. So the definition of the Covariant Derivative of $\mathbf{v}$ along $\mathbf{u}$ (which defines the direction) is:

$$\nabla_{\mathbf{u}}\mathbf{v} = \lim_{\epsilon \to 0} \left\{ \frac{[\mathbf{v}(\lambda_0 + \epsilon)]_{parallel\ transported\ to\ \lambda_0} - \mathbf{v}(\lambda_0)}{\epsilon} \right\} \qquad (76)$$

For the vector $\mathbf{v}$ to be parallel transported along the curve in the direction of $\mathbf{u}$, the tangent vector, we require that $\nabla_{\mathbf{u}}\mathbf{v} = 0$

When $\mathbf{u}$ is the tangent vector to a curve $P(\lambda)$, $\mathbf{u} = dP/d\lambda$, one uses the notation $DT_\alpha^\beta/d\lambda$ for the components of $\nabla_{\mathbf{u}}\mathbf{T}$.

From [13]:

Although the method of presentation may be quite different from the original, the following ideas are essentially those of the Italian mathematician Tullio Levi-Civita. Suppose we restrict our consideration to a curve C in Euclidean space. Let $V^j = V^j(\lambda)$ be components of a vector field of parallel vectors of constant magnitude along a curve C (parallel in the Euclidean sense); then

$$\frac{dV^j}{d\lambda} = 0. \qquad (77)$$

A definition of a parallel vector field in a Riemannian space is introduced only with respect to a given curve C of that space. The analytic form of the concept is inferred directly from (77).

Chapter 4

Geodesics

Definition   Let $\mathbf{V}$ be a vector field along a curve in C in a Riemannian space. The vector field is said to be a parallel vector field with respect to the curve C if and only if

$$\frac{DV^j}{d\lambda} = 0. \qquad (78)$$

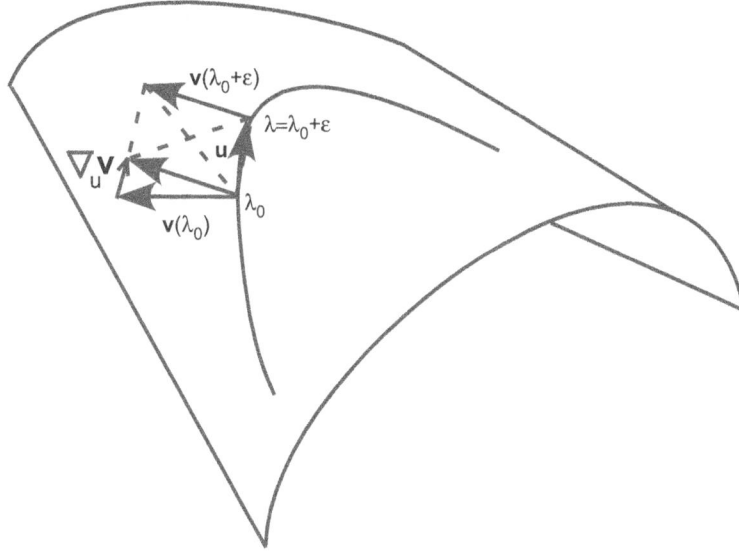

Figure 7: Covariant Derivative of **v** along **u**

We define the DIRECTIONAL COVARIANT DERIVATIVE to be[2]:

$$\frac{D}{d\lambda} = \frac{dx^i}{d\lambda}\nabla_i \tag{79}$$

We define PARALLEL TRANSPORT of a vector V^i along the path $x^i(\lambda)$ to be the requirement that the covariant derivative of V^i along the path vanishes:

$$\frac{D}{d\lambda}\mathbf{V} = \frac{dx^i}{d\lambda}\nabla_i V^j = 0 \tag{80}$$

Define the tangent vector as T^a, then

$$T^a\nabla_a V^b = 0 \tag{81}$$

Note a curve that parallel transports its tangent vector is a geodesic. Let the tangent vector be T^a:

$$T^a\nabla_a T^b = 0 \tag{82}$$

defines a geodesic curve.

We will also use the notation $\nabla_\mathbf{u}\mathbf{T}$ for $\frac{D}{d\lambda}T^i_j$ for Tensor T^i_j.

Theorem Let **V** be a parallel vector field with respect to a curve C. The magnitude of **V** is constant.

Theorem Let **A** and **B** be a parallel vector fields along a curve C. The angle determined by **A** and **B** is constant along C.

22

Geodesic   From the above theorem lets consider the definition of a geodesic. From [12]:

> Intuitively, geodesics are lines that "curve as little as possible"; they are the "straightest possible lines" one can draw in a curved geometry. Given a derivative operator, ∇_a, we define a *geodesic* to be a curve whose tangent vector is parallel propagated along itself when parallel propagated along itself, i.e. a curve whose tangent, T^a satisfies the equation

$$T^a \nabla_a T^b = 0 \tag{83}$$

Definition   Suppose that a curve $\mathbf{x}^j = \mathbf{x}^j(t)$ joins points P_0 and P_1 then

$$s = \int_{t_0}^{t_1} f \, dt, \quad f = \left(\left| g_{jk} \frac{d\mathbf{x}^j}{dt} \frac{d\mathbf{x}^k}{dt} \right| \right)^{\frac{1}{2}} \tag{84}$$

is said to be the distance from P_0 to P_1 along the given curve.

For example in rectangular coordinates, $g_{jk} = \delta_{jk}$, and $f = \frac{\sqrt{|(dx^1)^2 + (dx^2)^2 + (dx^3)^2|}}{|dt|}$ and s is clearly the arc length along the curve $\mathbf{x}^j = \mathbf{x}^j(t)$.

Theorem   For a curve C joining P_0 and P_1 to be minimum length it is necessary that the parametric equations satisfy the Euler-Lagrange equations

$$\frac{\partial f}{\partial \mathbf{x}^j} - \frac{\partial f/\partial \dot{\mathbf{x}}^j}{dt} = 0. \tag{85}$$

where

$$f = (|g_{jk} \dot{\mathbf{x}}^j \dot{\mathbf{x}}^k|)^{\frac{1}{2}} \tag{86}$$

$$\dot{\mathbf{x}}^j = \frac{d\mathbf{x}^j}{dt} \tag{87}$$

Theorem   If f is given by:

$$f = (|g_{jk} \dot{\mathbf{x}}^j \dot{\mathbf{x}}^k|)^{\frac{1}{2}} \quad f \neq 0 \tag{88}$$

the Euler-Lagrange differential equations are equivalent to the set

$$\ddot{\mathbf{x}}^j + \Gamma_{pq}^j \dot{\mathbf{x}}^p \dot{\mathbf{x}}^q = \frac{d(\ln f)}{dt} \dot{\mathbf{x}}^j \tag{89}$$

Furthermore, if the parameter t represents arc length s, then the above reduces to

$$\frac{d^2 \mathbf{x}^j}{ds^2} + \Gamma_{pq}^j \dot{\mathbf{x}}^p \dot{\mathbf{x}}^q = 0 \tag{90}$$

where

$$\dot{\mathbf{x}}^j = \frac{d\mathbf{x}^j}{ds} \tag{91}$$

23

Definition The curves satisfying the differential equation in (90) is said to be a geodesic of the space.

Since the geodesic differential equations is second order, a unique solution is determined at a point P_0 when x_0^j and $\dot{x}_0^j$ are given; that is, a point and a direction uniquely determine a geodesic at a given point.

Simplifications when g_{ij} is diagonal

$$\Gamma_{ij}^i = \Gamma_{ji}^i = \frac{\partial}{\partial x^j}\left(\frac{1}{2}ln|g_{ii}|\right) \tag{92}$$

$$\Gamma_{jj}^i = -\frac{1}{2g_{ii}}\partial_i g_{jj} \quad (i \neq j) \tag{93}$$

$$\textit{All other } \Gamma_{jk}^i \textit{ vanish} \tag{94}$$

Geodesic Example

One way to compute the Christoffel coefficients Γ_{pq}^j is to derive the geodesic equation (90) using variation (Lagrangian) from which we can read off the various coefficients.

Example in Polar Coordinates First lets work with polar coordinates [2]. In this case,

$$ds^2 = dr^2 + r^2 d\theta^2 \tag{95}$$

The none zero components of the inverse metric are $g^{rr} = 1$ and $g^{\theta\theta} = r^{-2}$. The connection coefficients are,

$$\Gamma_{rr}^r = \frac{1}{2}g^{r\rho}(\partial_r g_{r\rho} + \partial_r g_{\rho r} - \partial_\rho g_{rr}) =$$

$$= \frac{1}{2}g^{rr}(\partial_r g_{rr} + \partial_r g_{rr} - \partial_r g_{rr}) +$$

$$\frac{1}{2}g^{r\theta}(\partial_r g_{r\theta} + \partial_r g_{\theta r} - \partial_\theta g_{rr})$$

$$= \frac{1}{2}(1)(0 + 0 - 0) + \frac{1}{2}(0)(0 + 0 - 0)$$

$$= 0.$$

$$\Gamma_{\theta\theta}^r = \frac{1}{2}g^{r\rho}(\partial_\theta g_{\theta\rho} + \partial_\theta g_{\rho\theta} - \partial_\rho g_{\theta\theta})$$

$$= \frac{1}{2}g^{rr}(\partial_\theta g_{\theta r} + \partial_\theta g_{rr} - \partial_r g_{rr}) - \partial_r g_{\theta\theta})$$

$$= \frac{1}{2}(1)(0 + 0 - 2r)$$

$$= -r.$$

24

Continuing,

$$\Gamma^r_{\theta r} = \Gamma^r_{r\theta} = 0$$
$$\Gamma^\theta_{rr} = 0$$
$$\Gamma^\theta_{r\theta} = \Gamma^\theta_{\theta r} = \frac{1}{r}$$
$$\Gamma^\theta_{\theta\theta} = 0.$$

Since we have two coordinates $\mathbf{e}_r$ and $\mathbf{e}_\theta$, the geodesic equations are:

$$\frac{d^2\theta}{ds^2} + \Gamma^\theta_{r\theta}\frac{dr}{ds}\frac{d\theta}{ds} + \Gamma^\theta_{\theta r}\frac{d\theta}{ds}\frac{dr}{ds} = 0$$
$$\frac{d^2\theta}{ds^2} + \frac{2}{r}\frac{dr}{ds}\frac{d\theta}{ds} = 0 \tag{96}$$

$$\frac{d^2r}{ds^2} + \Gamma^r_{\theta\theta}\frac{d\theta}{ds}\frac{dr}{ds} = 0$$
$$\frac{d^2r}{ds^2} - r(\frac{d\theta}{ds}) = 0 \tag{97}$$

It can be shown that the solutions to the above two coupled differential equations is,

$$r = a\sec(\theta + b) \tag{98}$$

where a and b are constants. These are equations of a straight line confirming that the geodesics on flat space are straight lines. The shortest distance between two points in flat space is a straight line.

Definition of Geodesic [3]

A geodesic is a curve $P(\lambda)$ that parallel-transports its tangent vector $\mathbf{u} = dP/d\lambda$ along itself –

$$\nabla_\mathbf{u}\mathbf{u} = 0 \tag{99}$$

Notation

Note the new notation. This is equavalent to:

$$T^a\nabla_a T^b = 0 \tag{100}$$

where T^a is the tangent vector $\mathbf{u}$. We will use both notations. Also the notation for the covariant derivative of $\mathbf{v}$ in the direction $\mathbf{u}$ (or the vector V^b in the direction of the tangent vector U^b) is

$$\nabla_\mathbf{u}\mathbf{v} = U^b\nabla_b V^a \tag{101}$$

We will introduce even more notation later.

Geodesic from the Calculus of Variations

From [5]:

Theorem The integral $\int_a^b F(x, y, y')dx$, whose end points are fixed, is stationary for weak variations if y satisfies the differential equation

$$\frac{\partial F}{\partial y} - \frac{d}{dx}\left(\frac{\partial F}{\partial y'}\right) = 0 \tag{102}$$

where

$$y' = \frac{dy}{dx} \tag{103}$$

Also [5],

Theorem The integral $\int_a^b F(x, y')dx$, whose end points are fixed, is stationary for weak variations if y satisfies the differential equation

$$\frac{\partial F}{\partial y'} = c \tag{104}$$

where c is an arbitrary constant.

Geodesics on a Sphere For a sphere,

$$ds^2 = dx^2 + dy^2 + dz^2 = a^2 d\theta^2 + a^2 \sin^2 \theta d\phi^2 \tag{105}$$

where a is the radius.

So we are required to minimize the integral (note we are integrating over θ),

$$I = a \int_A^B \sqrt{1 + (\frac{d\phi}{d\theta})\sin^2 \theta} d\theta \tag{106}$$

From the theorem, $F(\theta, \frac{d\phi}{d\theta}) = \sqrt{1 + (\frac{d\phi}{d\theta})\sin^2 \theta}$ is a function of θ and $\frac{d\phi}{d\theta}$ which leads to

$$\frac{\partial}{\partial \frac{d\phi}{d\theta}} \sqrt{1 + (\frac{d\phi}{d\theta})sin^2\theta} = constant \tag{107}$$

$$\frac{\frac{d\phi}{d\theta}\sin^2\theta}{\sqrt{1 + (\frac{d\phi}{d\theta})\sin^2\theta}} = \sin\alpha \tag{108}$$

Solving for $\frac{d\phi}{d\theta}$ and integrating we get

$$\phi + \beta = \int \frac{\sin\alpha d\theta}{\sin\theta(\sin^2\theta - \sin^2\alpha)^{\frac{1}{2}}} \tag{109}$$

26

On substituting $\theta = \tan^{-1}(1/u)$ in the integral on the right-hand side it reduces to

$$-\int \frac{\tan\alpha\, du}{(1 - u^2 \tan^2\alpha)^{\frac{1}{2}}} \tag{110}$$

We can now integrate to obtain,

$$\cos(\phi + \beta) = \frac{\tan\alpha}{\tan\theta} \tag{111}$$

which, on transformation into Cartesian coordinates, gives us

$$x\cos\beta - y\sin\beta = z\tan\alpha \tag{112}$$

This is the equation of a plane through the center of the sphere.

"Thus the geodesics on a sphere are obtained as the intersection of the sphere and a plane through its center, and so must be arcs of great circles[5]."

More Notations for Covariant Derivative of Vector

$$u^\alpha_{;\beta} = u^\alpha_{,\beta} + \Gamma^\alpha_{\gamma\beta} u^\gamma \tag{113}$$

Where

$$u^\alpha_{,\beta} = \frac{\partial}{\partial x^\beta}\left(u^\alpha\right) \tag{114}$$

Thus, ";" emphasizes that corrections for "twist and turns" need to be taken into account when performing differentation.

Note in that in Cartesian Orthogonal Coordinate systems, $\Gamma^\alpha_{\gamma\beta} = 0$ and

$$u^\alpha_{;\beta} = u^\alpha_{,\beta} \tag{115}$$

Differential Equation for Geodesic from Definition:

$$\nabla_{\mathbf{u}}\mathbf{u} = 0 \tag{116}$$

Note that $\mathbf{u}$ is a tangent vector along a curve $P(\lambda)$ which will turn out to be the Geodesic. Thus,

$$\mathbf{u} = \frac{dx^\alpha}{d\lambda} \tag{117}$$

$$u^\alpha_{;\beta} u^\beta = (u^\alpha_{,\beta} + \Gamma^\alpha_{\gamma\beta} u^\gamma) u^\beta = 0 \tag{118}$$

$$\frac{\partial}{\partial x^\beta}\left(\frac{dx^\alpha}{d\lambda}\right)\frac{dx^\beta}{d\lambda} + \Gamma^\alpha_{\gamma\beta}\frac{dx^\gamma}{d\lambda}\frac{dx^\beta}{d\lambda} = 0 \tag{119}$$

$$\frac{d^2 x^\mu}{d\lambda^2} + \Gamma^\alpha_{\gamma\beta}\frac{dx^\gamma}{d\lambda}\frac{dx^\beta}{d\lambda} = 0 \tag{120}$$

27

Chapter 5

1-Forms and Connection Coefficients

1-Forms Reviewed

See [3] for details. Consider again a scalar function $\phi(x, y, z)$. Then the gradient is:

$$\nabla \phi = \frac{\partial \phi}{\partial x} \mathbf{i} + \frac{\partial \phi}{\partial y} \mathbf{j} + \frac{\partial \phi}{\partial z} \mathbf{k} \tag{121}$$

In terms of the basis vectors $\mathbf{e}_i$,

$$\nabla \phi = \frac{\partial \phi}{\partial x} \mathbf{e_1} + \frac{\partial \phi}{\partial y} \mathbf{e_2} + \frac{\partial \phi}{\partial z} \mathbf{e_3} \tag{122}$$

Or,

$$\nabla \phi = \frac{\partial \phi}{\partial x^i} \mathbf{e_i} \tag{123}$$

Now the differential of the scalar function $\phi(x, y, z)$ or $\phi(\mathbf{x}^i)$ is:

$$\mathbf{d}\phi = \frac{\partial \phi}{\partial x} \mathbf{d}x + \frac{\partial \phi}{\partial y} \mathbf{d}y + \frac{\partial \phi}{\partial z} \mathbf{d}z \tag{124}$$

$$\mathbf{d}\phi = \frac{\partial \phi}{\partial \mathbf{x}^i} d\mathbf{x}^i \tag{125}$$

Define,

$$\omega^i = d\mathbf{x}^i \tag{126}$$

Then,

$$\mathbf{d}\phi = \omega^i \frac{\partial \phi}{\partial \mathbf{x}^i} \tag{127}$$

Just as $\mathbf{e}_i$ are the basis coordinates, ω^i can also be viewed as basis coordinates in a dual space.

Now lets generalize. Define a dual vector,

$$\mathbf{U}_i = \omega^i U_i \tag{128}$$

Note that the vector $\mathbf{U_i}$ uses lower index as apposed to $\mathbf{V}^i = V^i \mathbf{e}_i$. Now,

$$\mathbf{U}_i \mathbf{V}^i = U_i V^i \tag{129}$$

which is a Real number. *1-forms map vectors into real numbers.* This implies,

$$\omega^i \mathbf{e}_j = \delta^i_j \tag{130}$$

Lets analyze this equation (130). As we noted $\omega^i = d\mathbf{x}^i$. So ω^i are the differences along the basis $\mathbf{e}_i$ for each x^i. So e_1 pierces planes of constant x^1 where each difference between planes is related to ω^1. So $\omega^1 e^1$ pierces only one plane (count one) by definition and e^1 does not pierce ω^2 and ω^3 or constant planes in the direction of e_2 and e_3. See Figure 8

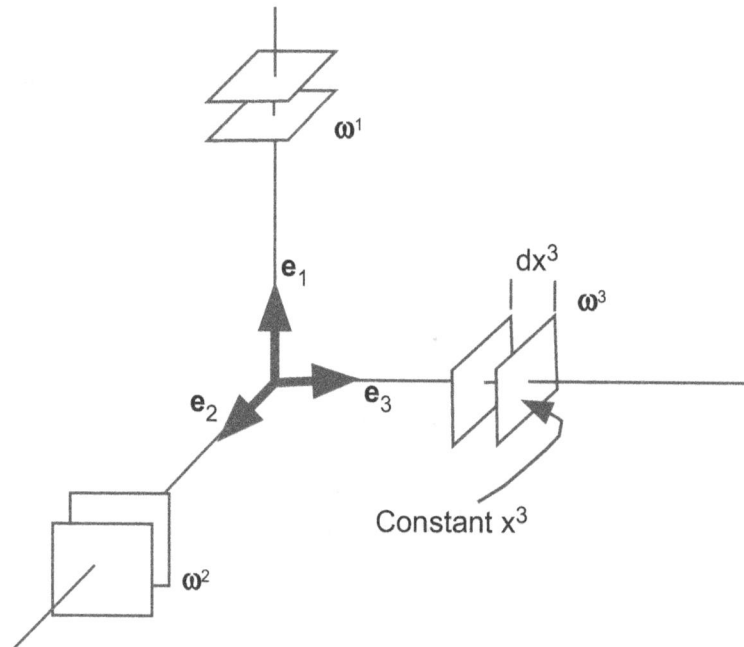

Figure 8: Basis vectors and 1-forms

Mathematical definition of 1-form: *a 1-form is a linear, real valued function of vectors.* Define the operation $<,>$ where we plug in a dual vector $\mathbf{u}^*$ and vector $\mathbf{v}$ then

$$< \mathbf{u}^*, \mathbf{v} > \tag{131}$$

yields a real number.

A single physical quantity can be described equally well by a vector $\mathbf{u}$ or by the corresponding 1-form $\mathbf{u}^*$.

29

Box 2.3 on Differentials from Gravitation [3]:

The "exterior drivative" or "gradient" $\mathbf{d}f$ of a function f is a more rigorous version of the elementary concept of "differential."

In elementary textbooks, one is presented with the differential df as representing "an infinitesimal change in the function $f(P)$" associated with some infinitesimal displacement of the point P; but one will recall that the displacement of P is left arbitrary, albeit infinitesimal. Thus df represents a change in f in some unspecified direction.

But this is precisely what the exterior derivative $\mathbf{d}f$ represents. Choose a particular, infinitesimally long displacement $\mathbf{v}$ of the point P. Let the displacement vector $\mathbf{v}$ pierce $\mathbf{d}f$ to give the number $< \mathbf{d}f, \mathbf{v} > = \partial_\mathbf{v} f$. That number is the change of f in going from the tail of $\mathbf{v}$ to its tip. Thus $\mathbf{d}f$, before it has been pierced to give a number, represents the change in f in an unspecified direction. The act of piercing $\mathbf{d}f$ with $\mathbf{v}$ is the act of making explicit the direction in which change is to be measured. The only failing of the textbook presentation, then, was its suggestion that $\mathbf{d}f$ was a scalar or a number; the explicit recognition of the need for specifying a direction $\mathbf{v}$ to reduce $\mathbf{d}f$ to a number $< \mathbf{d}f, \mathbf{v} >$ shows that in fact $\mathbf{d}f$ is a 1-form, the gradient of f.

Connection Coefficients   The following is from Gravitation [3].

To work with components, one needs a set of basis vectors $(\mathbf{e}_\alpha)$ and the dual set of basis 1-forms $\{\omega^\alpha\}$. In flat spacetime a single such basis suffices; all events can use the same Lorentz basis. Not so in curved spacetime! There each event has its own tangent space, and each tangent space requires a basis of its own. As one travels from event to event, comparing their bases via parallel transport, one sees the basis twist and turn. They must do so. In no other way can they accomidate themselves to the curvature of spacetime. Bases at points P_0 and P_1, which are the same when compared by parallel transport along one curve, must differ when compared along another curve.

To quantify the twisting and turning of a "field" of basis vectors $\{\mathbf{e}_\alpha(P)\}$ and forms $\{\omega^\alpha(P)\}$, use the covariant derivative. Examine the changes in vector fields along a basis vector $\mathbf{e}_\beta$, abbreviating

$$\nabla_{\mathbf{e}_\beta} \equiv \nabla_\beta \quad (definition\ of\ \nabla_\beta); \tag{132}$$

and especially examine the rate of change of some basis vector: $\nabla_\beta \mathbf{e}_\alpha$. This rate of change is itself a vector, so it can be expanded in terms of the basis:

$$\nabla_\beta \mathbf{e}_\alpha = \mathbf{e}_\mu \Gamma^\mu_{\alpha\beta} \tag{133}$$

Note reversal of order of α and β!

and the resultant "connection coefficients" $\Gamma^\mu_{\alpha\beta}$ can be calculated by projection on the basis 1-forms:

$$< \omega_\mu, \nabla \mathbf{e}_\alpha >= \Gamma^\mu_{\alpha\beta} \tag{134}$$

Because the basis 1-forms are "locked into" the basis vectors $(< \omega^\beta, \mathbf{e}_\alpha >= \delta^\beta_\alpha)$, these same connection coefficients $\Gamma^\mu_{\alpha\beta}$ tell how the 1-form basis changes from point to point:

$$\nabla_\beta \omega^\mu = -\Gamma^\mu_{\alpha\beta} \omega^\alpha, \tag{135}$$

$$< \nabla_\beta \omega^\mu, \mathbf{e}_\alpha >= -\Gamma^\mu_{\alpha\beta}, \tag{136}$$

Chapter 6

Curvature

Commutators

In the following $\mathbf{u} = \partial_\mathbf{u}$ and $\mathbf{v} = \partial_\mathbf{v}$. Other equivalent notation: $\mathbf{u} = \nabla_\mathbf{u}$. So they all represent covariant derivative along the direction $\mathbf{u}$ or $\mathbf{v}$.

From Gravitation [3]:

> At each point P a vector field $\mathbf{u}$ provides a vector $\mathbf{u}(P)$ –which is a differential operator $\partial_{\mathbf{u}(P)}$ – at each point P in some region of spacetime. This vector field operates on a function f to produce not just a number, but another function $\mathbf{u}[f] \equiv \partial_\mathbf{u}$. A second order vector field $\mathbf{v}$ can perfectly well operate on this new function, to produce yet another function
>
> $$\mathbf{v}\{\mathbf{u}[f]\} = \partial_\mathbf{v}(\partial_\mathbf{u} f) \tag{137}$$

Does this function agree with the result of applying $\mathbf{v}$ first and then $\mathbf{u}$? Equivalently, does the "commutator"

31

$$[\mathbf{u}, \mathbf{v}][f] \equiv \mathbf{u}\{\mathbf{v}[f]\} - \mathbf{v}\{\mathbf{u}[f]\} \qquad (138)$$

vanish? The simplest special case is when $\mathbf{u}$ and $\mathbf{v}$ are basis vectors of a coordinate system, $\mathbf{u} = \partial/\partial x^\alpha, \mathbf{v} = \partial/\partial x^\beta$. Then the commutator does vanish, because partial derivatives always commute:

$$[\partial/\partial x^\alpha, \partial/\partial x^\beta][f] = \partial^2 f/\partial x^\beta \partial x^\alpha - \partial^2 f/\partial x^\alpha \partial x^\beta = 0 \qquad (139)$$

But in general the commutator is nonzero, as one sees from a coordinate-based calculation:

$$[\mathbf{u}, \mathbf{v}]f = u^\alpha \frac{\partial}{\partial x^\alpha}\left(v^\beta \frac{f}{\partial x^\beta}\right) - v^\alpha \frac{\partial}{\partial x^\alpha}\left(u^\beta \frac{f}{\partial x^\beta}\right)$$
$$= \left[(u^\alpha v^\beta_{,\alpha} - v^\alpha u^\beta_{,\alpha})\frac{\partial}{\partial x^\beta}\right]f. \qquad (140)$$

Notice however, that the commutator $[\mathbf{u}, \mathbf{v}]$, like $\mathbf{u}$ and $\mathbf{v}$ themselves, is a vector field., i.e., a linear differential operator at each point event:

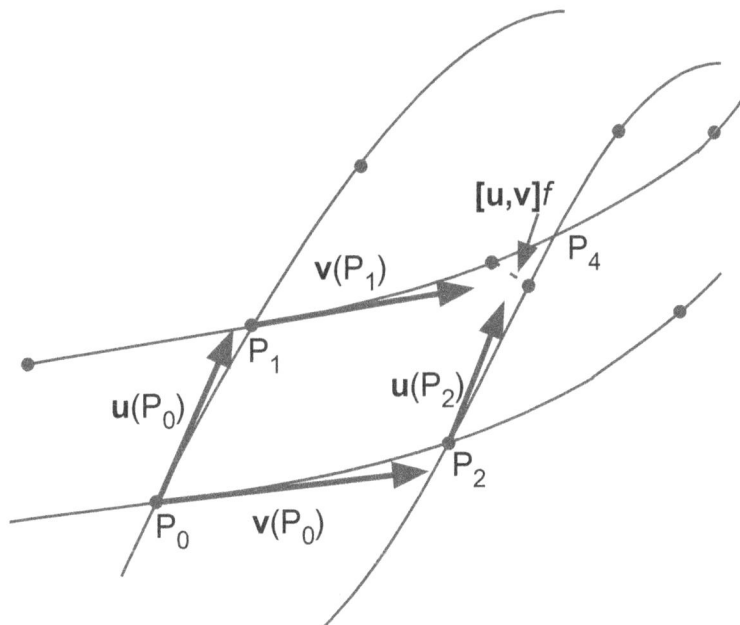

Figure 9: Commutators

$$[\mathbf{u}, \mathbf{v}] = (\mathbf{u}[v^\beta] - \mathbf{v}[u^\beta])\frac{\partial}{\partial x^\beta} = (u^\alpha v^\beta_{,\alpha} - v^\alpha u^\beta_{,\alpha})\frac{\partial}{\partial x^\beta}. \qquad (141)$$

32

Recall the notation:

$$f_{,\alpha} = \partial f / \partial x^\alpha \qquad (142)$$

Relative Acceleration of Neighboring Geodesics

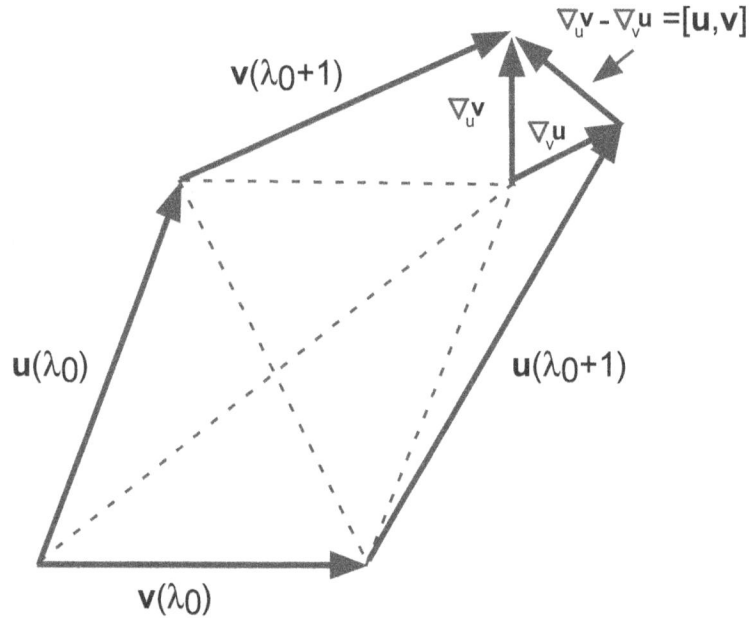

Figure 10: Commutators Illustrated

From Gravitation [3]:

> Take two vector fields. Combine into one the two diagrams for
> $\nabla_{\mathbf{u}}\mathbf{v}$ and $\nabla_{\mathbf{v}}\mathbf{u}$. See Figure 10. Thereby discover that $\nabla_{\mathbf{u}}\mathbf{v} - \nabla_{\mathbf{v}}\mathbf{u}$
> is the vector by which the $\mathbf{v} - \mathbf{u} - \mathbf{v} - \mathbf{u}$ quadrilateral fails to close
> – i.e. it is the commutator $[\mathbf{u}, \mathbf{v}] : \nabla_{\mathbf{u}}\mathbf{v} - \nabla_{\mathbf{v}}\mathbf{u} = [\mathbf{u}, \mathbf{v}]$.

Curvature Operator

In Figure 11 the vector $\mathbf{A}$ is parallel transported around a closed curve. The change in the vector as it completes the transport around the closed curve is $\delta\mathbf{A}$. In the diagram note the $\mathbf{u} = \frac{\partial}{\partial \mathbf{u}} = \nabla_{\mathbf{u}}$ and $\mathbf{v} = \frac{\partial}{\partial \mathbf{v}} = \nabla_{\mathbf{u}}$ are operators that act on the vector giving the covariant derivative in the respective directions (change in vector compared to parallel transport along path).

First change in parallel transport in $\mathbf{u}$ direction.

$$\delta_1 = \nabla_{\mathbf{u}}\mathbf{A}\Delta a \qquad (143)$$

33

Second change in transporting in **v** direction,

$$\delta_2 = \nabla_{\mathbf{v}} \delta_1 \Delta b = \nabla_{\mathbf{u}} \nabla_{\mathbf{v}} \mathbf{A} \Delta a \Delta b \tag{144}$$

Now transport **A** in direction of commutator,

$$\delta_3 = \nabla_{[\mathbf{v},\mathbf{u}]} \mathbf{A} \Delta a \Delta b \tag{145}$$

Transport along **v** in reverse direction,

$$\delta_4 = -\nabla_{\mathbf{v}} \mathbf{A} \Delta b \tag{146}$$

Transport along **u** in reverse direction,

$$\delta_5 = -\nabla_{\mathbf{v}} \delta_4 \Delta b \Delta a = -\nabla_{\mathbf{v}} \nabla_{\mathbf{u}} \mathbf{A} \Delta a \Delta b \tag{147}$$

This closes the curve. Now add up,

$$-\delta \mathbf{A} = \{\nabla_{\mathbf{u}} \nabla_{\mathbf{v}} - \nabla_{\mathbf{v}} \nabla_{\mathbf{u}} + \nabla_{[\mathbf{v},\mathbf{u}]}\} \mathbf{A} \Delta a \Delta b \tag{148}$$

Define the curvature operator,

$$R(\mathbf{u}, \mathbf{v}) = [\nabla_{\mathbf{u}}, \nabla_{\mathbf{v}}] - \nabla_{[\mathbf{u},\mathbf{v}]} \tag{149}$$

Finally,

$$\frac{-\delta \mathbf{A}}{\Delta a \Delta b} = R(\mathbf{u}, \mathbf{v}) \mathbf{A} \tag{150}$$

$$R^{\alpha}_{\beta\gamma\delta} \equiv \text{Riemann}(\omega^{\alpha}, \mathbf{e}_{\beta}, \mathbf{e}_{\gamma}, \mathbf{e}_{\delta}) \equiv <\omega^{\alpha}, R(\mathbf{e}_{\gamma}, \mathbf{e}_{\delta})\mathbf{e}_{\beta}> \tag{151}$$

For component basis $\{\mathbf{e}_{\alpha}\} = \partial/\partial x^{\alpha}$

$$R^{\alpha}_{\beta\gamma\delta} = \frac{\partial \Gamma^{\alpha}_{\beta\delta}}{\partial x^{\gamma}} - \frac{\partial \Gamma^{\alpha}_{\beta\gamma}}{\partial x^{\delta}} + \Gamma^{\alpha}_{\mu\gamma}\Gamma^{\mu}_{\beta\delta} - \Gamma^{\alpha}_{\mu\delta}\Gamma^{\mu}_{\beta\gamma} \tag{152}$$

The arrow δP in Figure 14 measures the second derivative:

$$(first\ derivative\ at\ \lambda + \frac{1}{2}\Delta\lambda) = \nabla_{\mathbf{n}} = \frac{M_p R - M_p B}{\Delta\lambda \Delta n} = \frac{BR}{\Delta\lambda \Delta n} \tag{153}$$

$$(first\ derivative\ at\ \lambda - \frac{1}{2}\Delta\lambda) = \nabla_{\mathbf{n}} = \frac{M_n Q - M_n A}{\Delta\lambda \Delta n} = \frac{-AQ}{\Delta\lambda \Delta n} \tag{154}$$

Transpose to common location λ, take difference, and divide it by $\Delta\lambda$ to obtain the second covariant derivative with respect to the vector **u**; thus

$$\nabla_{\mathbf{u}} \nabla_{\mathbf{u}} \mathbf{n} = \frac{(\nabla_{\mathbf{u}})_{\lambda + \frac{1}{2}\Delta\lambda} - (\nabla_{\mathbf{u}})_{\lambda - \frac{1}{2}\Delta\lambda}}{\Delta\lambda}$$

$$= \frac{(BR + AQ)_{vectors\ transported\ to\ common\ location}}{(\Delta\lambda)^2 \Delta n} = \frac{\delta_P}{(\Delta\lambda)^2 \Delta n} \tag{155}$$

Which equals the "relative acceleration vector" for neighboring geodesics.

34

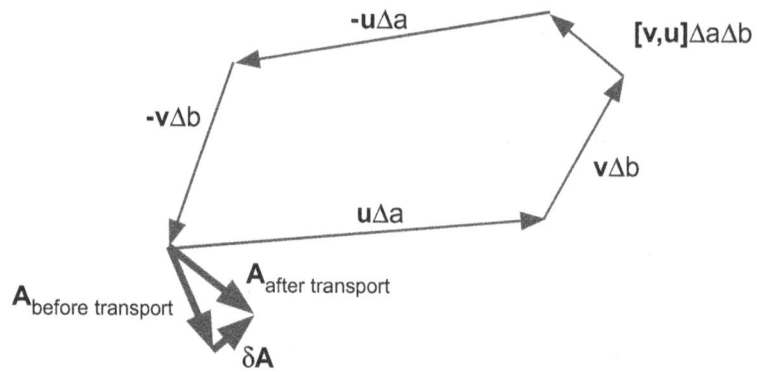

Figure 11: Closed Curve Parallel Transport

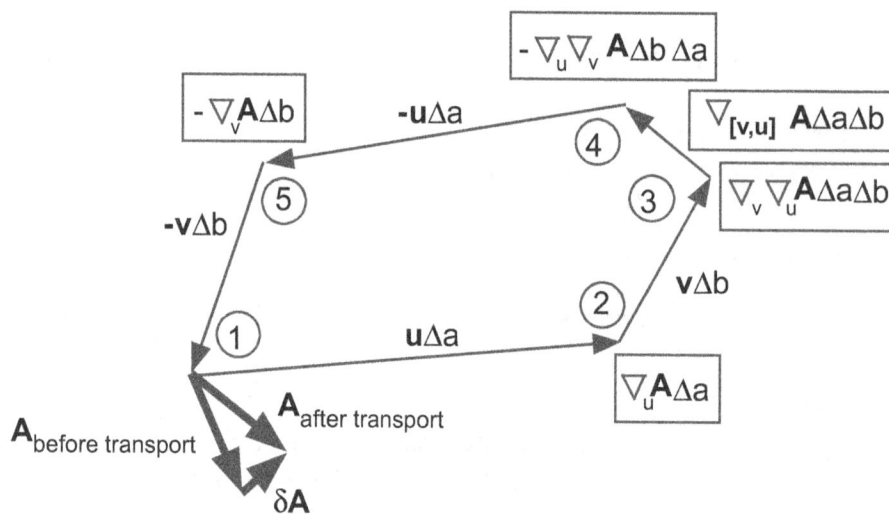

Figure 12: Closed Curve Details Showing Changes per Leg

35

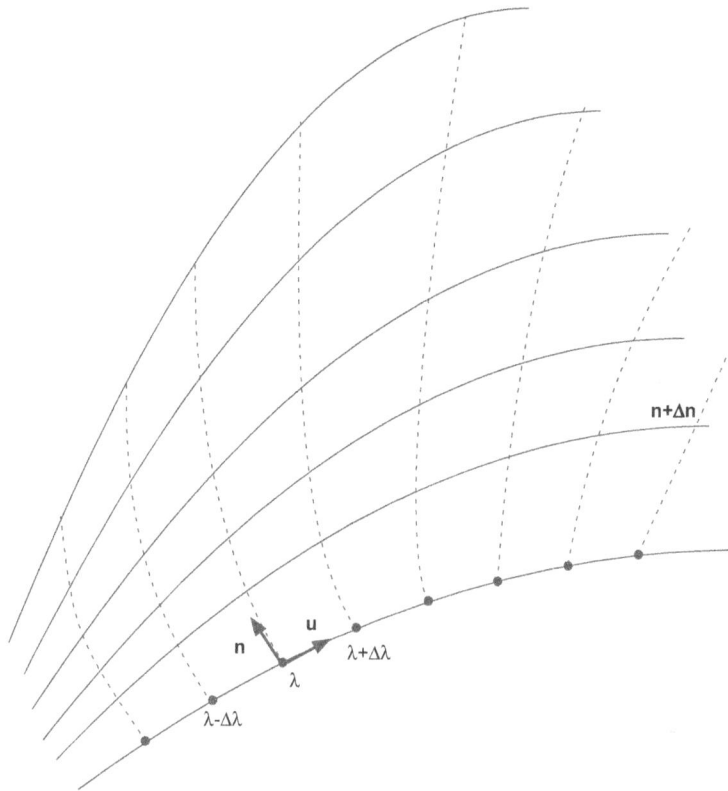

Figure 13: One Parameter Family of Geodesics

Equation of Geodesic Deviation

The equation of geodesic deviation is:

$$\nabla_{\mathbf{u}}\nabla_{\mathbf{u}}\mathbf{n} + R(\mathbf{n}, \mathbf{u})\mathbf{u} = 0 \tag{156}$$

which states that the change of a vector as a result of parallel transport around the loop $\mathbf{u}$ and $\mathbf{n}$ is equal to the change due to geodesic acceleration $\nabla_{\mathbf{u}}\nabla_{\mathbf{u}}\mathbf{n}$. Refer to Box 11.6 in Gravitation [3].

Derivation of the Equation of Geodesic Deviation Now lets derive the equation of geodesic deviation directly by computing $\nabla_{\mathbf{u}}\nabla_{\mathbf{u}}\mathbf{n}$. See Wald [12].

36

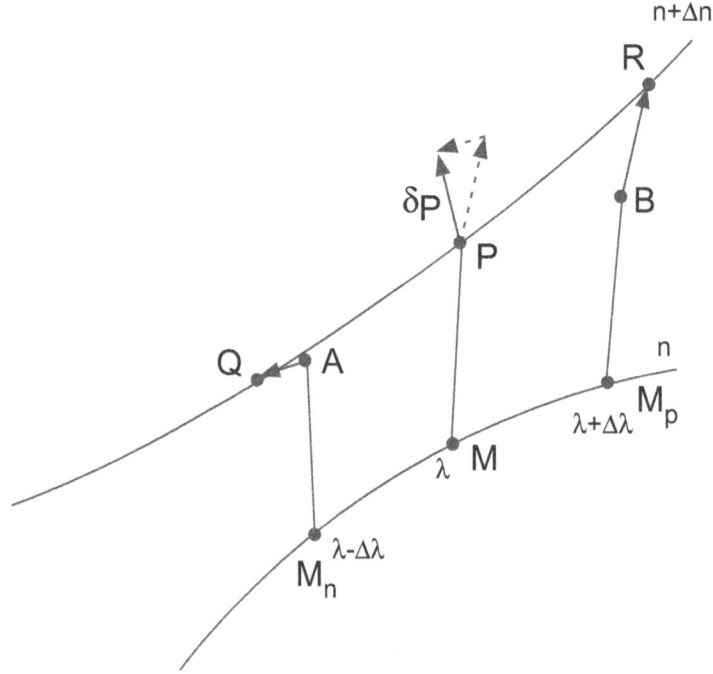

Figure 14: Geodesic Deviation

Denote the tangent $\mathbf{u}$ by the vector U^c. Also denote the vector for $\mathbf{n}$ as N^a.

$$
\begin{aligned}
\nabla_{\mathbf{u}}\nabla_{\mathbf{u}}\mathbf{n} &= U^c\nabla_c(U^b\nabla_b N^a) \\
&= U^c\nabla_c(N^b\nabla_b U^a) \\
&= (U^c\nabla_c N^b)(\nabla_b U^a) + N^b U^c\nabla_c\nabla_b U^a \\
&= (N^c\nabla_c U^b)(\nabla_b U^a) + N^b U^c\nabla_b\nabla_c U^a - R^a_{cbd}N^b U^c U^d \\
&= N^c\nabla_c(U^b\nabla_b U^a) - R^a_{cbd}N^b U^c U^d \\
&= -R^a_{cbd}N^b U^c U^d
\end{aligned}
\tag{157}
$$

In deriving the above equation we note that since U^c and N^b are coordinate basis they commute. That is $U^b\nabla_b N^a = N^b\nabla_b U^a$.

Noting that,

$$
R(\mathbf{n},\mathbf{u}) = [\nabla_{\mathbf{n}},\nabla_{\mathbf{u}}] - \nabla_{[\mathbf{n},\mathbf{u}]}
\tag{158}
$$

and

$$
\nabla_{[\mathbf{n},\mathbf{u}]} = 0
\tag{159}
$$

since $\mathbf{u}$ and $\mathbf{n}$ are coordinate basis vectors and commute.

$$
\nabla_{\mathbf{n}}\mathbf{u} - \nabla_{\mathbf{u}}\mathbf{n} = \left[\frac{\partial}{\partial n},\frac{\partial}{\partial \lambda}\right] = \frac{\partial^2}{\partial n\partial \lambda} - \frac{\partial^2}{\partial \lambda\partial n} = 0
\tag{160}
$$

37

Then

$$[\nabla_b, \nabla_c]U^a = R^a_{cbd}U^d \tag{161}$$

Chapter 7

Special Theory of Relativity

7.1 Spacetime

One of the postulates of the special theory of relativity is that in all inertial frames regardless of uniform velocity with respect to other frames the speed of light is constant. A photon travels at the speed of $c = 3 \times 10^8$ m/s regardless of the speed of the object in vacuum. In three dimensions the following holds true for a photon as it travels from the origin at time 0:

$$c^2 t^2 = x^2 + y^2 + z^2 \tag{162}$$

Or at any location,

$$c^2 (dt)^2 = (dx)^2 + (dy)^2 + (dz)^2 \tag{163}$$

Let us select units such that $c = 1$. Define ds such that,

$$(ds)^2 = -(dt)^2 + (dx)^2 + (dy)^2 + (dz)^2 \tag{164}$$

Let us define $x^\mu = [t, x, y, z] = [x^0, x^1, x^2, x^3] = [x^0, x^i]$ where $i = 1, 2, 3$ refers to space coordinates (Latin indexes). When we use Greek symbols for the index we refer to spacetime coordinates. We can write,

$$(ds)^2 = -(dx^0)^2 + (dx^1)^2 + (dx^2)^2 + (dx^3)^2 \tag{165}$$

For a particle moving at the speed of light, $ds = 0$. For particles moving at less than the speed of light, $(ds)^2 < 0$. We refer to these particles as moving in timelike manner. If a particle moves faster than the speed of light then $(ds)^2 > 0$. These are spacelike particles. If we treat $(ds)^2$ as a metric then we have,

$$(ds)^2 = dx^\alpha dx^\beta \eta_{\alpha\beta} \tag{166}$$

The metric $\eta_{\alpha\beta}$ is a (0,2) tensor. In matrix notation,

$$\eta_{\alpha\beta} = \begin{pmatrix} -1 & 0 & 0 & 0 \\ 0 & 1 & 0 & 0 \\ 0 & 0 & 1 & 0 \\ 0 & 0 & 0 & 1 \end{pmatrix} \tag{167}$$

The key point is that $(ds)^2$ is an invariant in any coordinate system. So let us consider a barred coordinate system, $\bar{x}^\mu$. Then,

$$(ds)^2 = d\bar{x}^\alpha d\bar{x}^\beta \eta_{\alpha\beta} \tag{168}$$

Let,

$$\bar{x}^\mu = \Lambda^\mu_\beta x^\beta \tag{169}$$

Where Λ^β_μ is a $(1,1)$ tensor. It is a 4×4 matrix. Then,

$$(ds)^2 = \Lambda^\mu_\alpha dx^\alpha \eta_{\mu\gamma} \Lambda^\gamma_\beta dx^\beta \tag{170}$$

Whence,

$$\eta_{\alpha\beta} = \Lambda^\mu_\alpha \eta_{\mu\gamma} \Lambda^\gamma_\beta \tag{171}$$

The matrices Λ^μ_α that satisfy (171) are known as the **Lorentz transformations**. The metric (167) is called **Lorentzian**.

The inverse metric $\eta^{\mu\gamma}$ is defined by:

$$\eta^{\mu\gamma} \eta_{\mu\gamma} = \delta^\mu_\nu \tag{172}$$

It has identical components to $\eta_{\mu\gamma}$. Let

$$x^\mu = \bar{\Lambda}^\mu_\beta \bar{x}^\beta \tag{173}$$

Then we can show that

$$\bar{\Lambda}^\mu_\beta \Lambda^\beta_\rho = \delta^\mu_\rho \tag{174}$$

We can lower indices,

$$A_\mu = \eta_{\mu\nu} A^\nu \tag{175}$$

Consider a boost in which coordinates are changed to a frame that travels at a constant velocity. Let both coordinate systems coincide at t=0. Further let the boost be in the x^1 direction with no velocity in x^2 and x^3. That is, $\bar{x}^2 = x^2$, $\bar{x}^3 = x^3$. Then,

$$\bar{x}^0 = \lambda^0_0 x^0 + \lambda^0_1 x^1 \tag{176}$$

$$\bar{x}^1 = \lambda^1_0 x^0 + \lambda^1_1 x^1 \tag{177}$$

Where,

$$\Lambda^\alpha_\beta = \begin{pmatrix} \lambda^0_0 & \lambda^0_1 & 0 & 0 \\ \lambda^1_0 & \lambda^1_1 & 0 & 0 \\ 0 & 0 & 1 & 0 \\ 0 & 0 & 0 & 1 \end{pmatrix} \tag{178}$$

From the orthogonality relationship (167)

$$\begin{aligned} -1 &= -(\lambda^0_0)^2 + (\lambda^0_1)^2 \\ 1 &= -(\lambda^0_1)^2 + (\lambda^1_1)^2 \\ 0 &= -\lambda^0_1 \lambda^0_0 + \lambda^1_1 \lambda^1_0 \end{aligned} \tag{179}$$

40

Close examination of (179) and comparison with the relationship between hyperbolic functions leads to:

$$
\begin{aligned}
\lambda_0^0 &= \cosh\phi \\
\lambda_1^0 &= -\sinh\phi \\
\lambda_0^1 &= -\sinh\phi \\
\lambda_1^1 &= \cosh\phi
\end{aligned}
\tag{180}
$$

$$
\Lambda_\beta^\alpha =
\begin{pmatrix}
\cosh\phi & -\sinh\phi & 0 & 0 \\
-\sinh\phi & \cosh\phi & 0 & 0 \\
0 & 0 & 1 & 0 \\
0 & 0 & 0 & 1
\end{pmatrix}
\tag{181}
$$

$$
\begin{aligned}
\bar{x}^0 &= x^0 \cosh\phi - x^1 \sinh\phi \\
\bar{x}^1 &= -x^0 \sinh\phi + x^1 \cosh\phi
\end{aligned}
\tag{182}
$$

$$
\begin{aligned}
\bar{t} &= t \cosh\phi - x \sinh\phi \\
\bar{x} &= -t \sinh\phi + x \cosh\phi
\end{aligned}
\tag{183}
$$

The point defined by $\bar{x} = 0$ is moving with velocity v.

$$
v = \frac{x}{t} = \frac{\sinh\phi}{\cosh\phi} = \tanh\phi
\tag{184}
$$

Replace ϕ with $\tanh^{-1} v$ in (183) to obtain,

$$
\begin{aligned}
\bar{t} &= \gamma(t - vx) \\
\bar{x} &= \gamma(x - vt)
\end{aligned}
\tag{185}
$$

where $\gamma = 1/\sqrt{1 - v^2}$.

Theorem Lorentz Contraction [13] Let $\bar{L}$ be a rod fixed in an $\bar{O}$ coordinate system and in uniform rectilinear motion in relation to an O system. The length measurement of $\bar{L}$ is shorter when determined from the viewpoint of the O system than it is when determined by $\bar{O}$ measurements. In particular,

$$
\bar{L}_0 = \sqrt{1 - v^2}\,\bar{L}_{\bar{O}}
\tag{186}
$$

Theorem Time Dilation [13] Let I be a time interval associated with an $\bar{O}$ coordinate system, which is in uniform rectilinear motion with respect to an O system. The O determination of this time interval is then less than the $\bar{O}$ measurement. We have

$$\bar{I}_0 = \sqrt{1 - v^2}\bar{I}_{\bar{O}} \tag{187}$$

Given two 4-vectors A^μ and B^β, then the inner product is:

$$A^\mu \eta_{\mu\nu} B^\nu = A^\mu B_\mu \tag{188}$$

A non-zero 4- vector A^μ is called **spacelike** if $A^\mu A_\mu > 0$, **timelike** if $A^\mu A_\mu < 0$, and **null** if $A^\mu A_\mu = 0$.

7.2 World-Lines and Proper Time

This section draws on [11]. Let $x^\mu(\lambda)$ be a parameterized curve in Minkowski space with parameter λ defined over an interval $a \geq \lambda \leq b$. Also assume that in the interval x^μ is **timelike**. Define the **tangent 4-vector** to the curve at an event p to be:

$$U = U^\mu e_\mu \ where \ U^\mu = \frac{dx^\mu}{d\lambda}\mid_{\lambda = \lambda_0} \tag{189}$$

Note that the tangent vector U^μ is independent of the choice of orthonormal basis e_μ. The path of a material particle will be assumed to be timelike at all events through which it passes. This path is referred to as the particle's **world-line**. This assumption amounts to the requirement that the particle's velocity is always less than c.

$$U^\mu U_\mu = \eta_{\mu\nu}\frac{dx^\mu(\lambda)}{d\lambda}\frac{dx^\nu(\lambda)}{d\lambda} = (\frac{dt}{d\lambda})^2\left(\sum_{i=1}^{3}(\frac{dx^i(t)}{dt})^2 - 1\right) < 0 \tag{190}$$

In the above, $t = x^0 = t(\lambda)$. Hence,

$$v^2 = \sum_{i=1}^{3}(\frac{dx^i(t)}{dt})^2 < 1 \tag{191}$$

In conventional dimensions, this means that $v^2 < c^2$. For two neighboring events on the world-line, $x^\mu(\lambda)$ and $x^\mu(\lambda + \Delta\lambda)$, set

$$\Delta\tau^2 = -\Delta s^2 = -\eta_{\mu\nu}\Delta x^\mu \Delta x^\nu > 0, \tag{192}$$

where,

$$\Delta x^\mu = \Delta x^\mu(\lambda + \Delta\lambda) - \Delta x^\mu(\lambda) \tag{193}$$

In the limit $\Delta\lambda \to 0$

$$\Delta\tau^2 \to -\eta_{\mu\nu}\frac{dx^\mu(\lambda)}{d\lambda}\frac{dx^\nu(\lambda)}{d\lambda}(\Delta\lambda)^2 = -(v^2 - 1)(\frac{dt}{d\lambda})^2(\Delta\lambda)^2 \tag{194}$$

42

Hence

$$\Delta\tau \rightarrow \sqrt{1 - v^2}\Delta t = \frac{1}{\gamma}\Delta t \qquad (195)$$

Note that $\Delta\tau$ is time dilated with respect to t. Hence, τ is the time as measured by a clock moving with the particle in the world-line. Measure the length from p to q:

$$\tau_{pq} = \int_p^q d\tau = \int_{t_p}^{t_q} \frac{dt}{\gamma} \qquad (196)$$

From [11]:

> Define τ as the **proper time** from p to q. If the event p is fixed and we let q vary along the curve then the proper time can be used as a parameter along the curve,
>
> $$\tau = \int_{t_p}^t \frac{dt}{\gamma} = \tau(t) \qquad (197)$$
>
> The tangent 4-vector $V = V^\mu e_\mu$ calculated with respect to this special parameter is called the **4-velocity** of the particle,
>
> $$V^\mu = \frac{dx^\mu}{d\tau} = \gamma\frac{dx^\mu}{dt} \qquad (198)$$

Form the inner product,

$$V^\mu V_\mu = (-1 + v^2)\gamma^2 = -(1 - v^2)\gamma^2 = -1 \qquad (199)$$

Thus, the magnitude of the 4-velocity is a constant and is invariant under coordinate transformation.

Define the **4-acceleration** of a particle as $A = A^\mu e_\mu$ with components

$$A^\mu = \frac{dV^\mu}{d\tau} = \frac{d^2 x^\mu(\tau)}{d\tau^2} \qquad (200)$$

The components expressed in terms of the time parameter t are

$$A^\mu = \gamma\left(\frac{d\gamma}{dt} \, , \, \frac{d\gamma}{dt}\mathbf{v} + \gamma\frac{d\mathbf{v}}{dt}\right) \qquad (201)$$

where

$$\mathbf{v} = \sum_{i=1}^3 \frac{dx^i}{dt}e_i \qquad (202)$$

The 4-velocity and 4-acceleration are orthogonal.

This is consequence of (199). If we take the derivative of (199)with respect to proper time we get,

$$V^\mu V_\mu = -1$$
$$d(V^\mu V_\mu)/d\tau = 0$$
$$2V^\mu\frac{dV_\mu}{d\tau} = 0$$
$$V^\mu A_\mu = 0 \qquad (203)$$

43

7.3 Energy and Momentum

Let the particle's rest mass be $m = m_0$. Define the **4-momentum**

$$p^\mu = mV^\mu \tag{204}$$

Now,

$$p^\mu p_\mu = -m^2 \tag{205}$$

So the length of the 4-momentum vector is an invariant. For example under a boost, $\bar{p}^\mu \bar{p}_\mu = -m^2$. Consider a boost in the x direction. Then,

$$\bar{p}^\mu = (\gamma m, v\gamma m, 0, 0) \tag{206}$$

For small v we have $\bar{p}^0 = m + \frac{1}{2}mv^2$. Thus, define $E = \bar{p}^0$ as the energy. If we re-introduce c, we have,

$$E = mc^2 + \frac{1}{2}mv^2 \tag{207}$$

Hence, the energy of the moving particle has increased by $\frac{1}{2}mv^2$ the Kinetic energy. Note that the total energy includes the rest mass energy. Thus, $E = mc^2$ for a particle at rest. Although we will not insist on this but we can also think that the mass of the moving particle has increased by the Kinetic energy. We would like to keep the rest mass as the invariant always refering to m.

Since we have associated p^0 with Energy, then for $\bar{p}^\mu \bar{p}_\mu = -m^2$ we have

$$E^2 = m^2 + \mathbf{p}^2 \tag{208}$$

where $\mathbf{p}^2 = \delta_i^j p^i p^j$.

Definition   The quantity,

$$m = \frac{m_0}{\sqrt{(1 - v^2)}} = \gamma m_0 \tag{209}$$

is called the inertial mass (m_0 is the rest mass).

The above definition shows that there is a limit to the maximum speed of a particle with mass. The photon which travels at the speed of light has zero rest mass.

Definition Lorentz contraction factor for volume is $(1 - v^2)^{-\frac{1}{2}}$.

Chapter 8

Energy Momentum Stress Tensor

In this chapter we will provide a review of Fluid Dynamics since we are interested in a Perfect Fluid in Spacetime.

For a comprehensive treatment of Fluid Dynamics see [1]. In the following, we will refer to the second-order tensor stress tensor τ_{ij} which is fully developed in [1].

At this point it is instructive to illustrate the *components of the stress tensor*, τ_{ij}. In Figure 15 a cubic volume element is shown. Let $\mathbf{T}_k$ denotes the stress over δS_k with $k = 1, 2, 3$. For each face k we can decompose the stress as:

$$\mathbf{T}_k = \tau_{k1}\mathbf{e_1} + \tau_{k2}\mathbf{e_2} + \tau_{k3}\mathbf{e_3} \tag{210}$$

where τ_{ki} represents the jth component $(j = 1, 2, 3)$ of the stress over the face δS_k. In [8], it is shown that the quantities $\tau_{jk}(j = 1, 2, 3; k = 1, 2, 3)$ constitute the components of an affine tensor of rank 2. The adjective *affine* is used in order to emphasize that the coordinate transformation referred to in this definition is an orthogonal one. Thus, for an affine transformation,

$$\bar{\tau}_{jk} = \sum_{h=1}^{3} \sum_{l=1}^{3} \frac{\partial \bar{x}^j}{\partial x^h} \frac{\partial \bar{x}^k}{\partial x^l} \tau_{hl} \tag{211}$$

and

$$\tau_{jk} = \sum_{h=1}^{3} \sum_{l=1}^{3} \frac{\partial x^j}{\partial \bar{x}^h} \frac{\partial x^k}{\partial \bar{x}^l} \bar{\tau}_{hl} \tag{212}$$

Divergence Theorem

Let $\mathbf{F}(x, y, z) = F_x(x, y, z)\mathbf{i} + F_y(x, y, z)\mathbf{j} + F_z(x, y, z)\mathbf{k}$ be an arbitrary vector function. For example, in a fluid the velocity $\mathbf{u}(x, y, z)$ is a vector function. See Figure 17 where the surface S completely encloses the volume V. Then the divergence theorem states that,

$$\iint_S \mathbf{F} \cdot \hat{n} ds = \iiint_V \nabla \cdot \mathbf{F} dV \tag{213}$$

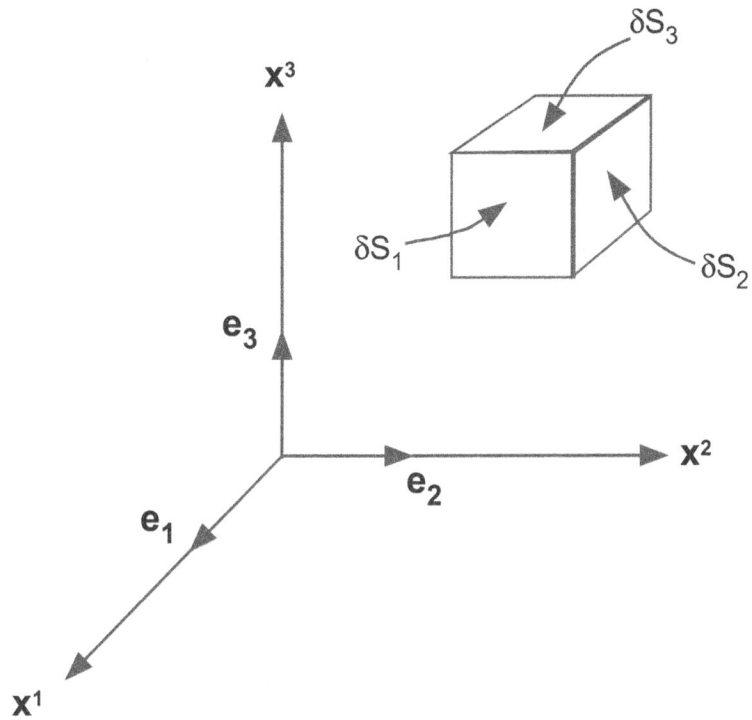

Figure 15: Stress Tensor Components Setup

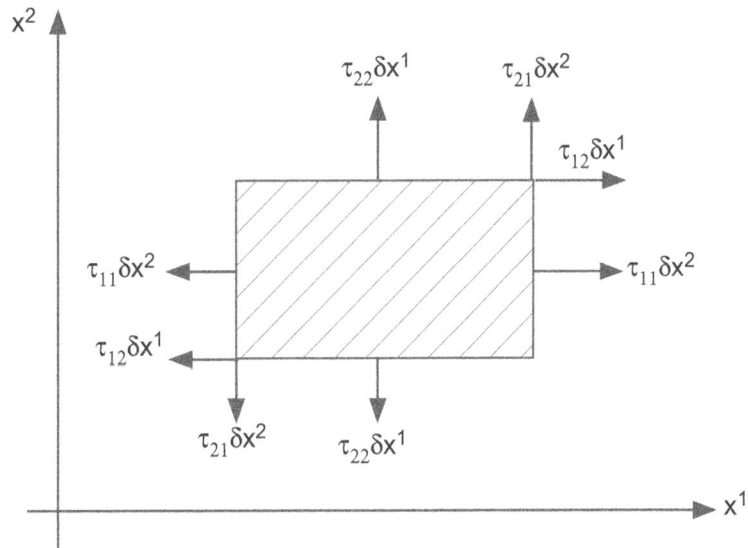

Figure 16: The surface forces acting on a rectangular element of fluid of unit depth

where,

$$\nabla \cdot \mathbf{F} = \frac{\partial F_x(x,y,z)}{\partial x} + \frac{\partial F_y(x,y,z)}{\partial y} + \frac{\partial F_z(x,y,z)}{\partial z} \tag{214}$$

and

$$\nabla = \frac{\partial}{\partial x}\mathbf{i} + \frac{\partial}{\partial y}\mathbf{j} + \frac{\partial}{\partial z}\mathbf{j} \tag{215}$$

Continuity Equation

See [10] for details. In Figure 17 the volume V at any time t contains

$$\iiint_V \rho(x,y,z,t)dV \tag{216}$$

amount of matter. In the above $\rho(x,y,z.dt)$ is the density of stuff at a given location at time t. The rate at which the stuff in the volume is changing is,

$$\frac{d}{dt}\iiint_V \rho(x,y,z,t)dV = \iiint_V \frac{\partial \rho}{\partial t}dV \tag{217}$$

Now lets compute the rate at which stuff enters and leaves (flows through) the surface S.

$$\iint_S \rho\mathbf{v} \cdot \hat{\mathbf{n}}dS \tag{218}$$

where $\mathbf{v}$ is the velocity of the flowing stuff and $\hat{\mathbf{n}}$ is the unit normal to the surface at position x,y,z. We assert that the rate at which the amount of stuff in V is changing is equal to the rate at which the stuff is flowing through the surface S that encloses V [10] . Thus,

$$\iiint_V \frac{\partial \rho}{\partial t}dV = -\iint_S \rho\mathbf{v} \cdot \hat{\mathbf{n}}dV \tag{219}$$

Apply the divergence theorem (213):

$$\iint_S \rho\mathbf{v} \cdot \hat{\mathbf{n}}dS = \iiint_V \nabla \cdot (\rho\mathbf{v})dV \tag{220}$$

Hence,

$$\iiint_V \frac{\partial \rho}{\partial t}dV = -\iiint_V \nabla \cdot (\rho\mathbf{v})dV \tag{221}$$

From which we arrive at the continuity equation,

$$\frac{\partial \rho}{\partial t} = -\nabla \cdot (\rho\mathbf{v}) \tag{222}$$

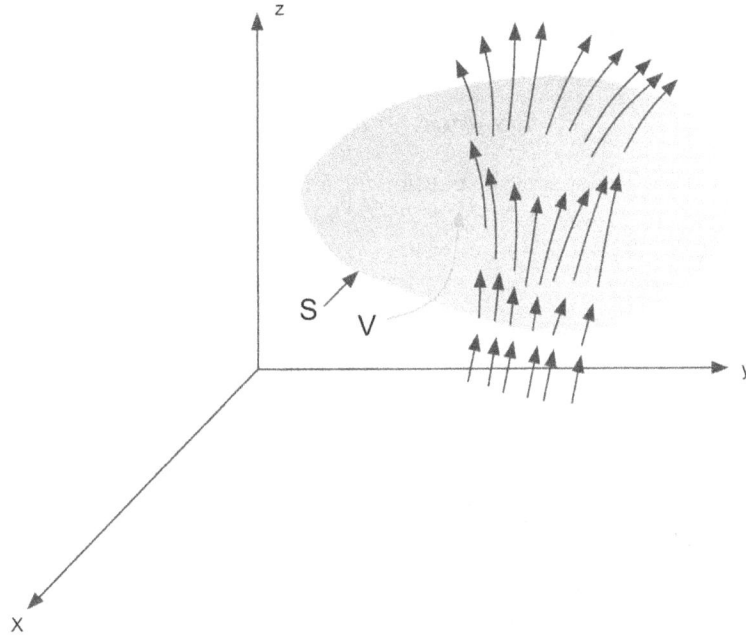

Figure 17: Divergence

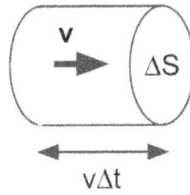

Figure 18: Flow Orthogonal

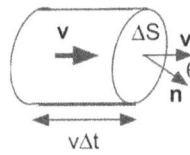

Figure 19: Flow Oblique

48

8.1 Differentiation Following the Motion of Fluid

For steady flow, the change in velocity of a material element is (note that the material element moves to $\delta t\mathbf{u}$),

$$\mathbf{u}(\mathbf{x} + \mathbf{u}\delta t, t + \delta t) - \mathbf{u}(\mathbf{x}, t) \tag{223}$$

Noting that the Taylor series expansion for a function of two variables is $f(x + \Delta x, y + \Delta y) = f(x, y) + \Delta x \frac{\partial f}{\partial x} + \Delta y \frac{\partial f}{\partial y} + \cdots$. For vectors, $f(\mathbf{x} + \Delta \mathbf{x}) = f(\mathbf{x}) + \Delta \mathbf{x} \cdot \nabla f(\mathbf{x}) + \cdots$ we can write,

$$\mathbf{u}(\mathbf{x} + \mathbf{u}\delta t, t + \delta t) - \mathbf{u}(\mathbf{x}, t) = \delta t \frac{\partial \mathbf{u}}{\partial t} + (\delta t\mathbf{u}) \cdot \nabla \mathbf{u} + O(\delta t^2) \tag{224}$$

From which we can introduce the notation,

$$\frac{D}{Dt} = \frac{\partial}{\partial t} + \mathbf{u} \cdot \nabla \tag{225}$$

$\frac{D}{Dt}$ is referred to as the material derivative.

The acceleration of a fluid element can be written $\frac{D\mathbf{u}}{Dt}$

Recall the conservation of mass equation:

$$\frac{\partial \rho}{\partial t} + \nabla \cdot (\rho \mathbf{u}) = 0 \tag{226}$$

expanding the divergence term $\nabla \cdot (\rho \mathbf{u}) = \rho(\nabla \mathbf{u}) + \mathbf{u} \cdot \nabla \rho$, we can write,

$$\frac{1}{\rho} \frac{D\rho}{Dt} + \nabla \cdot \mathbf{u} = 0 \tag{227}$$

Interpretation of $\nabla \cdot \mathbf{u}$. The volume V of a material body of fluid changes as a result of the movement of each element $\mathbf{n}\delta S$ of the bounding material surface. Thus,

$$\frac{dV}{dt} = \int \mathbf{u} \cdot \mathbf{n} dS = \int \nabla \cdot \mathbf{u} dV \tag{228}$$

where we have used the divergence theorem. From [1], "the rate at which the volume of a material element instantaneously enclosing the point $\mathbf{x}$ is changing, divided by the volume is,"

$$\lim_{V \to 0} \frac{1}{V} \frac{dV}{dt} = \lim_{V \to 0} \frac{1}{V} \int \nabla \cdot \mathbf{u} dV = \nabla \cdot \mathbf{u} \tag{229}$$

This fractional rate of change of the volume of a material element is called the (local) *rate of expansion* or rate of dilation [1].

49

Fluid Equation of Motion

Consider the body of fluid with volume V enclosed by surface S. The momentum is $\int \mathbf{u}\rho dV$. Force is the rate of change of momentum. So the rate of change of momentum is:

$$\int \frac{D\mathbf{u}}{Dt}\rho dV \tag{230}$$

which can be interpreted as the sum of the products of mass and acceleration through out the elements of the material volume V. This force is equal to the sum of the body forces and the surface or contact forces, see ([1]).

mass $\times$ acceleration = resultant of body forces + resultant of surface forces
$$\tag{231}$$

First consider the body force.

$$\mathbf{F} = \int \mathbf{f}\rho dV \tag{232}$$

The surface force on the area δS with normal $\mathbf{n}$ has a component in the i direction $\tau_{ij}n_j\delta S$ where the summation over j is implied. The ith component of this force over the area S enclosing the volume V is,

$$\int \tau_{ij}n_j dS = \int \frac{\partial \tau_{ij}}{\partial x_j} dV \tag{233}$$

where we have used the divergence theorem.

Writing down equation(231),

$$\int \frac{Du_i}{Dt}\rho dV = \int \mathbf{f}\rho dV + \int \frac{\partial \tau_{ij}}{\partial x_j} dV \tag{234}$$

The integral relation holds for all choices of the material volume V. So that,

$$\rho\frac{Du_i}{Dt} = \rho f_i + \frac{\partial \tau_{ij}}{\partial x^j} \tag{235}$$

This is the equation of motion and holds at all points in the fluid. In vector form,

$$\rho\frac{D\mathbf{u}}{Dt} = \rho\mathbf{f} + \nabla \cdot \mathbf{T} \tag{236}$$

Fluid at Rest

For a fluid at rest $\rho\frac{D\mathbf{u}}{Dt} = 0$ or,

$$\rho\mathbf{f} = -\nabla \cdot \mathbf{T} \tag{237}$$

Consider that the only force acting on the fluid volume is gravity. Then $\mathbf{f} = \mathbf{g}$ and,

50

$$\nabla \cdot \mathbf{T} = -\rho \mathbf{g} \qquad (238)$$

For a fluid at rest the tensor $\mathbf{T}$ is diagonal as the shear forces arise due to fluid motion. Let, $\mathbf{T} = -p\delta_{ij}$. In this case,

$$\frac{\partial p}{\partial x_3} = \rho g \qquad (239)$$

where (x, yz) corresponds to (x^1, x^2, x^3). Thus,

$$p = p_0 + \rho g z \qquad (240)$$

assuming that ρ is independent of z. For the fluid at rest in this case p is the pressure.

8.2   Stress Tensor

For a fluid at rest only normal stresses are exerted and the normal stress is independent of the direction of the normal to the surface element across which it acts. In this case,

$$\tau_{ij} = -p\delta_{ij} \qquad (241)$$

For a fluid in motion we can expect,

$$\tau_{ij} = -p\delta_{ij} + d_{ij} \qquad (242)$$

where the terms $d_{ij} = 0$ for a fluid at rest. Also p in(242) is not the pressure except for the case when the fluid is at rest. The non-isotropic part d_{ij} is termed the *deviatic stress tensor* and is entirely due to the existance of motion in the fluid.

For a fluid in motion it is expected that the velocity component u_i changes for an incremental increase in the direction x^j about a point. "The hypothesis is made that d_{ij} is approximately a linear function of the various components of the velocity gradient for sufficiently small magnitudes of those components. Analytically the hypothesis is expressed as [1],

$$d_{ij} = A_{ijkl}\frac{\partial u_k}{\partial x^l} \qquad (243)$$

where the fourth-order tensor coefficient A_{ijkl} depends on the local state of the fluid, but not directly on the velocity distribution, and is necessarily symmetrical in the indices i and j like d_{ij}."

To make a long story short the deviatoric stress tensor expression is [1]:

$$d_{ij} = 2\mu(e_{ij} - \frac{1}{3}\nabla \cdot \mathbf{u}\delta_{ij}) \qquad (244)$$

where e_{ij} is the symmetrical part of $\frac{\partial u_k}{\partial x^l}$.

The constant μ is the *viscosity* of the fluid.

Navier-Stokes Equation

Writing down the total stress tensor,

$$\tau_{ij} = -p\delta_{ij} + 2\mu(e_{ij} - \frac{1}{3}\nabla \cdot \mathbf{u}\delta_{ij}) \tag{245}$$

and substituting into the equation of motion (236),

$$\rho\frac{Du_i}{Dt} = \rho f_i - \frac{\partial p}{\partial x^i} + \frac{\partial}{\partial x^j}\left\{2\mu(e_{ij} - \frac{1}{3}\nabla \cdot \mathbf{u}\delta_{ij})\right\} \tag{246}$$

This is called the *Navier-Stokes equation of motion.*

If we assume that appreciable differences in temperature do not exist for the fluid (μ depends significantly on temperature), then μ is independent of position. Then,

$$\rho\frac{Du_i}{Dt} = \rho f_i - \frac{\partial p}{\partial x^i} + \mu\left(\frac{\partial^2 u_i}{\partial x^j \partial x^j} + \frac{1}{3}\frac{\partial}{\partial x^j}\nabla \cdot \mathbf{u}\delta_{ij}\right) \tag{247}$$

Recall the equation of conservation of mass,

$$\frac{1}{\rho}\frac{D\rho}{Dt} + \nabla \cdot \mathbf{u} = 0 \tag{248}$$

For an incompressible fluid, the mass conservation equation reduces to:

$$\nabla \cdot \mathbf{u} = 0 \tag{249}$$

and we get,

$$\rho\frac{D\mathbf{u}}{Dt} = \rho\mathbf{f} - \nabla p + \mu\nabla^2\mathbf{u} \tag{250}$$

Perfect Fluid

A perfect fluid is one for which d_{ij} vanish identically. Thus, from (242)

$$\nabla \cdot T = -\nabla p \tag{251}$$

In this case (236) reduces to

$$\rho[\partial_t\mathbf{v} + (\mathbf{v} \cdot \nabla)\mathbf{v}] = -\nabla p \tag{252}$$

where we have used the definition (225) . Also we assume no body force.

Newtonian Fluid

For fluids which,

$$d_{ij} = 2\mu(e_{ij} - \frac{1}{3}\nabla \cdot \mathbf{u}\delta_{ij}) \tag{253}$$

hold are called Newtonian fluids. For example, for a simple shearing motion, and taking $\frac{\partial u_1}{\partial x_2}$ as the one non-zero velocity derivative, all components of d_{ij} are zero except,

$$d_{12} = d_{21} = \mu \frac{\partial u_1}{\partial x^2} \tag{254}$$

"μ is the constant of proportionality between the rate of shear and the tangential force per unit area when plane layers of fluid slide over each other[1]". It is termed the *viscosity* of the fluid. This relationship was proposed by Newton.

8.3   Perfect Fluid in SpaceTime

According to [2]:

> Dust may be defined in flat spacetime as a collection of particles at rest with respect to each other. The four-velocity field $U^\mu(x)$ is clearly going to be the constant four-velocity of the individual particles. Indeed, its components will be the same at each point. Define the number-flux four-vector to be

$$N_\mu = nU_\mu \tag{255}$$

> where n is the number density of the particles as measured in their rest frame.

Now it is very important to take note of the invariant: "in any frame, the number density that would be measured if you were in the rest frame is a fixed quantity"[2].

If each particle has the same mass m then in the rest frame the energy density of the dust is given by

$$\rho = mn \tag{256}$$

Define the energy-momentum tensor for dust:

$$T_{dust}^{\mu\nu} = p^\mu N^\nu = mnU^\mu U^\nu = \rho U^\mu U^\nu \tag{257}$$

For a perfect fluid, $T^{11} = T^{22} = T^{33}$ and $\rho = T^{00}$. Furthermore, the pressure is $p = T^{ii}$. Note the pressure is equal in all directions. Also note that in the rest frame $p^\mu = (m, 0, 0, 0)$ and $N^\mu = (n, 0, 0, 0)$. The energy-momentum tensor of a perfect fluid takes the following form in its rest frame:

$$T^{\mu\nu} = \begin{pmatrix} \rho & 0 & 0 & 0 \\ 0 & p & 0 & 0 \\ 0 & 0 & p & 0 \\ 0 & 0 & 0 & p \end{pmatrix} \tag{258}$$

Referring to [2] the general form of the energy-momentum tensor for a perfect fluid is:

$$T^{\mu\nu} = (\rho + p)U^\mu U^\nu + p\eta^{\mu\nu} \tag{259}$$

We refer to [2] for the "somewhat arbitrary" way the equation is dervied but we note that there is complete confidence in the result. In particluar, "given that (258) should be the form of $T^{\mu\nu}$ in the rest frame, and that (259) is a perfectly tensorial expression that reduces to (258) in the rest frame, we know that (259) must be the right expression in any frame."

A property of the energy-moment tensor $T^{\mu\nu}$ is that it is conserved. This is a huge property indeed. Thus,

$$\partial_\mu T^{\mu\nu} = 0 \tag{260}$$

The expression is comprised of four equations one for each ν. For $\nu = 0$ the equations corresponds to the conservation of energy. The equations $\partial_\mu T^{\mu k}, k = 1, 2, 3$ correspond to conservation of the kth component of momentum.

If we take the nonrelativistic limit, where for $U^\mu = (1, v^i)$ we have $|v^i| << 1$ and $p << \rho$ where that last condition holds true since pressure comes about from the random motions of individual particles and these motions are small, then after some work [2] derives,

$$\rho[\partial_t \mathbf{v} + (\mathbf{v} \cdot \nabla)\mathbf{v}] = -\nabla p \tag{261}$$

from (260) which is the Euler equation in fluid mechanics. This is the same as we derived for the Perfect Fluid in (252).

The above is also derived in [3] page 152.

Chapter 9

Differential Forms

As we proceed, we will run into situations where formulation in terms of differential forms will provide great insight into the problem. The introduction given here is just enough to get an elementary understanding. We draw on Gravitation [3] and Lovelock and Rund [8] . It turns out that differential forms are also an alternative formulation in electromagnetics and offers better geometrical insight into fields [9]. So it is worth the effort especially when we cover Special Relativity and Electromagnetics. We have a full development in a later chapter. Construct the sum,

$$\omega = A_j dx^j \tag{262}$$

This sum is a number. However, we can introduce the function $dx_k(.)$ that assigns a vector its kth coordinate. Thus for a vector $\mathbf{A} = (A_1, ... A_n)$, we have $dx_k(\mathbf{A}) = A_k$.

We can introduce the function,

$$\omega_x(\mathbf{a}) = A_1(\mathbf{x}) dx_1(\mathbf{a}) + ... + A_n(\mathbf{x}) dx_n(\mathbf{a}) \tag{263}$$

Hence $\omega_x(\mathbf{a})$ produces a number from the vector $\mathbf{a}$. Think of the inner product between two vectors. Where, in this case , the operation on the vector by the 1-form produces the inner product. So the 1-form is like a machine in which you plug in a vector and you get back a number.

Some examples from [14],

1. If $\mathbf{a} = (-2, 0, 4)$ then $dx_1(\mathbf{a}) = -2, dx_2(\mathbf{a}) = 0, dx_3(\mathbf{a}) = 4$.

2- If in R^2, $\omega_x = \omega_{(x,y)} = x^2 dx + y^2 dy$, then $\omega_{(x,y)}(a, b) = ax^2 + by^2$ and $\omega_{(1,-3)}(a, b) = a + 9b$ producing a number.

Let,

$$\pi = B_j dx^j \tag{264}$$

Then,

$$\omega + \pi = (A_j + B_j) dx^j \tag{265}$$

which is a 1-form.

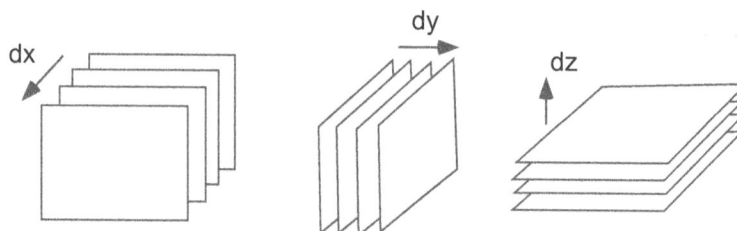

Figure 20: Fundmental 1-forms in Cartesian Coordinates

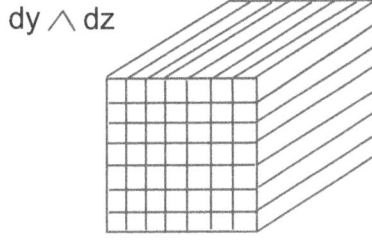

Figure 21: Fundmental 2-form $dy \wedge dz$ in Cartesian Coordinates

Exterior Product (Wedge Product)

As pointed out in [14], when the product of two 1-forms ω, π is considered, a new concept must be introduced. Define the wedge product,

$$\omega \wedge \pi = -\pi \wedge \omega \tag{266}$$

In particular,

$$dx^j \wedge dx^k = -dx^j \wedge dx^k \tag{267}$$

The exterior product of two 1-forms, ω and π can be expressed as,

$$\omega \wedge \pi = \frac{1}{2}(A_j B_k dx^j \wedge dx^k + A_j B_k dx^j \wedge dx^k) = \frac{1}{2}(A_j B_k - A_k B_j)dx^j \wedge dx^k \tag{268}$$

This new expression, noting the terms $dx^j \wedge dx^k$, is not a 1-form but a scalar 2-form. A general 2-form of this type is represented by expressions such as,

$$A_{jk}dx^j \wedge dx^k \tag{269}$$

Exterior Derivative of p-Forms Consider the 1-form: ω

$$\omega = A_i dx^i \tag{270}$$

define the exterior derivative, denoted by $d\omega$:

$$d\omega = \frac{\partial A_j}{\partial x^k} dx^k \wedge dx^j \tag{271}$$

which is a 2-form.

Using the fact that,

$$d\omega = -\frac{\partial A_j}{\partial x^k} dx^j \wedge dx^k \tag{272}$$

We can combine to write,

$$d\omega = -\frac{1}{2}\left(\frac{\partial A_k}{\partial x^j} - \frac{\partial A_j}{\partial x^k}\right)dx^j \wedge dx^k \tag{273}$$

56

This shows that "the basis elements $dx^j \wedge dx^k$ with $j < k$ are the components of the "curl" of the vector field A_j." [8].

The exterior derivative of the 2-form:

$$A_{jh} dx^j \wedge dx^h \tag{274}$$

is defined by:

$$d\omega = \frac{\partial A_{jh}}{\partial x^k} dx^k \wedge dx^j \wedge dx^h = \frac{\partial A_{jh}}{\partial x^k} dx^j \wedge dx^h \wedge dx^k \tag{275}$$

In [8] the following is derived:

$$d(\omega \wedge \pi) = d\omega \wedge \pi + (-1)^P \omega \wedge d\pi \tag{276}$$

which represents the "product rule for the exterior derivative of the exterior product of a p-form ω with a q-form π [8]."

Consider the exterior derivative $d(d\omega)$ of the 1-form,

$$d\omega = \frac{\partial A_j}{\partial x^k} dx^k \wedge dx^j \tag{277}$$

Then,

$$d(d\omega) = 0 \tag{278}$$

by virtue of the symmetry of the second derivatives :

$$\frac{\partial^2 A_h}{\partial x^k \partial x^j} = \frac{\partial^2 A_h}{\partial x^j \partial x^k} \tag{279}$$

It is important to note that in 3-dimensional space, if $d\omega$ is a one form the relationship $d(d\omega) = 0$ corresponds to $\nabla \times (\nabla f) = 0$ from vector calculus. If $d\omega$ is a two form the relationship $d(d\omega) = 0$ corresponds to $\nabla \cdot (\nabla \mathbf{F}) = 0$.

Chapter 10

Electromagnetics, Differential Forms and Special Relativity

In the following we will use the invariance of the principal that there are no single magnetic poles in the rest frame and in the moving frame to derive key equations in Maxwell's equations. See [3]. We also illustrate differential forms in electromagnetics and derive two compact equations that encapsulate the whole of Maxwell's equations.

The Lorentz force in EM is:

$$\frac{d\mathbf{p}}{dt} = e(\mathbf{E} + \mathbf{v} \times \mathbf{B}) \tag{280}$$

Define the 4-momentum:

$$\mathbf{P} = [p^0, \mathbf{p}] \tag{281}$$

Calculate,

$$\frac{d\mathbf{p}}{d\tau} = \frac{1}{\sqrt{1-v^2}}\frac{d\mathbf{p}}{dt} = \frac{e}{\sqrt{1-v^2}}(\mathbf{E} + \mathbf{v} \times \mathbf{B}) = e(u^0\mathbf{E} + \mathbf{u} \times \mathbf{B}) \tag{282}$$

$$\frac{d\mathbf{p}}{d\tau} = e\mathbf{F}(\mathbf{u}) \tag{283}$$

where $\mathbf{u}$ is the particles 4-velocity.

Energy-Charge Law

$$\frac{dp^0}{d\tau} = \frac{1}{\sqrt{1-v^2}}\frac{dE}{dt} = \frac{1}{\sqrt{1-v^2}}e\mathbf{E} \cdot \mathbf{u} \tag{284}$$

Note that $e\mathbf{E} \cdot dx$ is the work done in moving by dx. Thus, $e\mathbf{E}\frac{dx}{dt} = e\mathbf{E} \cdot \mathbf{v}$ is the rate of change with only component in $\mathbf{v}$ direction contributing.

$$\frac{d\mathbf{P}}{d\tau} = eF^\alpha_\beta u^\beta = \frac{dP^\alpha}{d\tau} \tag{285}$$

$$F^\alpha_\beta = \begin{array}{c} \\ 0 \\ 1 \\ 2 \\ 3 \end{array}\begin{array}{cccc} 0 & 1 & 2 & 3 \\ \left(\begin{array}{cccc} 0 & E_x & E_y & E_z \\ E_x & 0 & B_z & -B_y \\ E_y & -B_z & 0 & B_x \\ E_z & B_y & -B_x & 0 \end{array}\right) \end{array} \tag{286}$$

$$F^\alpha_\beta u^\beta = \begin{pmatrix} u^1 E_x + u^2 E_y + u^3 E_z \\ E_x u^0 + B_z u^2 - B_y u^3 \\ E_y u^0 - B_y u^1 + B_x u^3 \\ E_z u^0 + B_y u^1 - B_x u^2 \end{pmatrix} = \begin{pmatrix} \frac{dE}{d\tau} \\ f_x \\ f_y \\ f_z \end{pmatrix} \tag{287}$$

$$F_{\alpha\beta} = \eta_{\alpha\beta}F^\alpha_\beta \tag{288}$$

$$F_{\beta\alpha} = \begin{array}{c} \\ 0 \\ 1 \\ 2 \\ 3 \end{array} \begin{array}{cccc} 0 & 1 & 2 & 3 \\ \begin{pmatrix} 0 & -E_x & -E_y & -E_z \\ E_x & 0 & B_z & -B_y \\ E_y & -B_z & 0 & B_x \\ E_z & B_y & -B_x & 0 \end{pmatrix} \end{array} \qquad (289)$$

Recall that a boost of frame of reference by velocity parameter α in $z - t$ Plane

$$\text{Velocity } \beta = \tanh\alpha \qquad (290)$$

$$\sinh\alpha = \frac{\beta}{\sqrt{1 - \beta^2}} \qquad (291)$$

$$\cosh\alpha = \frac{1}{\sqrt{1 - \beta^2}} = \gamma \qquad (292)$$

$$x^\mu = \Lambda^\mu_\nu \bar{x}^\nu \qquad (293)$$

$$\bar{x}^\mu = \bar{\Lambda}^\nu_\mu x^\mu \qquad (294)$$

and in matrix notation:

$$\|\Lambda^\mu_\nu\| = \begin{array}{c} \\ 0 \\ 1 \\ 2 \\ 3 \end{array} \begin{array}{cccc} 0 & 1 & 2 & 3 \\ \begin{pmatrix} \cosh\alpha & 0 & 0 & \sinh\alpha \\ 0 & 1 & 0 & 0 \\ 0 & 0 & 1 & 0 \\ \sinh\alpha & 0 & 0 & \cosh\alpha \end{pmatrix} \end{array} \qquad (295)$$

$$\|\bar{\Lambda}^\mu_\nu\| = \begin{array}{c} \\ 0 \\ 1 \\ 2 \\ 3 \end{array} \begin{array}{cccc} 0 & 1 & 2 & 3 \\ \begin{pmatrix} \cosh\alpha & 0 & 0 & -\sinh\alpha \\ 0 & 1 & 0 & 0 \\ 0 & 0 & 1 & 0 \\ -\sinh\alpha & 0 & 0 & \cosh\alpha \end{pmatrix} \end{array} \qquad (296)$$

with

$$\Lambda^\gamma_\nu \bar{\Lambda}^\nu_\mu = \delta^\gamma_\mu \qquad (297)$$

Since $F_{\mu\nu}$ is a tensor, relative to the Lorentz transformation, we have,

$$\bar{F}_{\alpha\beta} = F_{\mu\nu} \Lambda^\mu_\alpha \Lambda^\nu_\beta \qquad (298)$$

Let us write the above in matrix form with $\mathbf{\Lambda} = \|\Lambda^\mu_\nu\|$, then

$$\bar{\mathbf{F}} = \mathbf{\Lambda}^T \mathbf{F} \mathbf{\Lambda} \qquad (299)$$

$$\mathbf{F} = \begin{pmatrix} 0 & -E_x & -E_y & -E_z \\ E_x & 0 & B_z & -B_y \\ E_y & -B_z & 0 & B_x \\ E_z & B_y & -B_x & 0 \end{pmatrix} \qquad (300)$$

59

$$\bar{\mathbf{F}} = \begin{pmatrix} \cosh\alpha & 0 & 0 & -\sinh\alpha \\ 0 & 1 & 0 & 0 \\ 0 & 0 & 1 & 0 \\ -\sinh\alpha & 0 & 0 & \cosh\alpha \end{pmatrix} \begin{pmatrix} 0 & -E_x & -E_y & -E_z \\ E_x & 0 & B_z & -B_y \\ E_y & -B_z & 0 & B_x \\ E_z & B_y & -B_x & 0 \end{pmatrix} \begin{pmatrix} \cosh\alpha & 0 & 0 & -\sinh\alpha \\ 0 & 1 & 0 & 0 \\ 0 & 0 & 1 & 0 \\ -\sinh\alpha & 0 & 0 & \cosh\alpha \end{pmatrix}$$

$$(301)$$

Multiplying through terms cancel out and we use $\cosh^2\alpha - \sinh^2\alpha = 1$ we get,

$$\bar{\mathbf{F}} = \begin{pmatrix} 0 & -E_x\cosh\alpha - B_y\sinh\alpha & -E_y\cosh\alpha + B_x\sinh\alpha & E_z \\ E_x\cosh\alpha + B_y & 0 & B_z & -E_x\cosh\alpha - B_y \\ E_y\cosh\alpha - B_x\sinh\alpha & -B_z & 0 & -E_y\cosh\alpha + B_x\sinh\alpha \\ E_z & E_x\cosh\alpha + B_y & E_y\cosh\alpha - B_x\sinh\alpha & 0 \end{pmatrix}$$

$$(302)$$

Writing out,

$$\bar{\mathbf{F}} = \begin{pmatrix} 0 & -\bar{E}_x & -\bar{E}_y & -\bar{E}_z \\ \bar{E}_x & 0 & \bar{B}_z & -\bar{B}_y \\ \bar{E}_y & -\bar{B}_z & 0 & \bar{B}_x \\ \bar{E}_z & \bar{B}_y & -\bar{B}_x & 0 \end{pmatrix} \qquad (303)$$

Recalling that $\sinh\alpha = \frac{\beta}{\sqrt{1-\beta^2}}$ and $\cosh\alpha = \frac{1}{\sqrt{1-\beta^2}}$

$$\bar{E}_x = \frac{1}{\sqrt{1-\beta^2}}(E_x + \beta B_y) \qquad (304)$$

$$\bar{E}_y = \frac{1}{\sqrt{1-\beta^2}}(E_y - \beta B_x) \qquad (305)$$

$$\bar{E}_z = E_z \qquad (306)$$

$$\bar{B}_x = \frac{1}{\sqrt{1-\beta^2}}(-\beta E_y + B_x) \qquad (307)$$

$$\bar{B}_y = \frac{1}{\sqrt{1-\beta^2}}(\beta E_x + B_y) \qquad (308)$$

$$\bar{B}_z = B_z \qquad (309)$$

Now from $\bar{x}^\mu = \Lambda^\mu_\nu x^\nu$ we have,

$$\bar{t} = t\cosh\alpha - z\sinh\alpha$$
$$\bar{x} = x$$
$$\bar{y} = y$$
$$\bar{z} = -t\sinh\alpha + z\cosh\alpha \qquad (310)$$

Since we are working with orthogonal coordinate transformations, for any scalar function ϕ we have,

$$\frac{\partial\phi}{\partial\bar{x}^\mu} = \sum_{\beta=0}^{4} \frac{\partial x^\beta}{\partial\bar{x}^\mu}\frac{\partial\phi}{\partial x^\beta} \qquad (311)$$

60

$$\frac{\partial \phi}{\partial \bar{x}^3} = \frac{\partial \phi}{\partial x^0}\frac{\partial x^0}{\partial \bar{x}^3} + \frac{\partial \phi}{\partial x^3}\frac{\partial x^3}{\partial \bar{x}^3} \tag{312}$$

Since $x^3 = \sinh \alpha \bar{x}^0 + \cosh \alpha \bar{x}^3$ and $x^0 = \cosh \alpha \bar{x}^0 + \sinh \alpha \bar{x}^3$ and letting $t = x^0$ and $z = x^3$, then

$$\frac{\partial \phi}{\partial \bar{z}} = \cosh \alpha \frac{\partial \phi}{\partial z} + \sinh \alpha \frac{\partial \phi}{\partial t} \tag{313}$$

Let $\beta << 1$ then,

$$\frac{\partial}{\partial \bar{z}} = \frac{\partial}{\partial z} + \beta \frac{\partial}{\partial t} \tag{314}$$

Now comes the invariance of physical laws with Lorentz transformation. In the rest frame,

$$\nabla \cdot \mathbf{B} = \frac{\partial B_x}{\partial x} + \frac{\partial B_y}{\partial y} + \frac{\partial B_z}{\partial z} = 0 \tag{315}$$

That is there is no free magnetic pole. Now this should also be true in the moving frame,

$$\nabla \cdot \bar{\mathbf{B}} = \frac{\partial \bar{B}_x}{\partial x} + \frac{\partial \bar{B}_y}{\partial y} + \frac{\partial \bar{B}_z}{\partial z} = 0 \tag{316}$$

We have,

$$\begin{aligned}
\frac{\partial}{\partial \bar{z}} &= \frac{\partial}{\partial z} + \beta \frac{\partial}{\partial t} \\
\frac{\partial}{\partial \bar{x}} &= \frac{\partial}{\partial x} \\
\frac{\partial}{\partial \bar{y}} &= \frac{\partial}{\partial y}
\end{aligned} \tag{317}$$

$$\frac{\partial B_z}{\partial z} + \beta \frac{\partial B_z}{\partial t} + \frac{\partial(-\beta E_y + B_x)}{\partial x} + \frac{\partial(\beta E_x + B_y)}{\partial y} = 0 \tag{318}$$

$$\beta \left(\frac{\partial B_z}{\partial t} - \frac{\partial E_y}{\partial x} + \frac{\partial E_x}{\partial y} \right) + \frac{\partial B_z}{\partial z} + \frac{\partial B_x}{\partial x} + \frac{\partial B_y}{\partial y} = 0 \tag{319}$$

$$\frac{\partial B_z}{\partial t} - \frac{\partial E_y}{\partial x} + \frac{\partial E_x}{\partial y} = 0 \tag{320}$$

In a similar development for direction along x or y we obtain,

$$\frac{\partial \mathbf{B}}{\partial t} + \nabla \times \mathbf{E} = 0 \tag{321}$$

Equations (315) and (321) together can be encapsulated in the following,

$$F_{\alpha\beta,\gamma} + F_{\beta\gamma,\alpha} + F_{\gamma\alpha,\beta} = 0 \tag{322}$$

61

with $\alpha = 1, \beta = 2, \gamma = 3$ etc. leading to $\nabla \cdot \mathbf{B} = 0$ and $\alpha = 0, \beta = 0, \gamma = 0$ leading to $\frac{\partial \mathbf{B}}{\partial t} + \nabla \times \mathbf{E} = 0$ whence,

$$\mathbf{dF} = 0 \tag{323}$$

and

$$\nabla \cdot *\mathbf{F} = 0 \tag{324}$$

General 2-form Faraday $\mathbf{F}$,

$$\mathbf{F} = \frac{1}{2} F_{\mu\nu} \mathbf{d}x^\mu \wedge \mathbf{d}x^\nu \tag{325}$$

Now take the exterior derivative of Faraday, using t, x, y, z for x^μ,

$$\begin{aligned}
\mathbf{dF} = &\left(\frac{\partial B_x}{\partial x} + \frac{\partial B_y}{\partial y} + \frac{\partial B_z}{\partial z} \right) \mathbf{d}x \wedge \mathbf{d}y \wedge \mathbf{d}z \\
+ &\left(\frac{\partial B_x}{\partial t} + \frac{\partial E_y}{\partial y} - \frac{\partial E_z}{\partial z} \right) \mathbf{d}t \wedge \mathbf{d}y \wedge \mathbf{d}z \\
+ &\left(\frac{\partial B_y}{\partial t} + \frac{\partial E_x}{\partial z} - \frac{\partial E_z}{\partial x} \right) \mathbf{d}t \wedge \mathbf{d}z \wedge \mathbf{d}x \\
+ &\left(\frac{\partial B_z}{\partial t} + \frac{\partial E_y}{\partial x} - \frac{\partial E_x}{\partial y} \right) \mathbf{d}t \wedge \mathbf{d}x \wedge \mathbf{d}y
\end{aligned} \tag{326}$$

Cleary since $\nabla \cdot \mathbf{B} = 0$ and $\frac{\partial \mathbf{B}}{\partial t} + \nabla \times \mathbf{E} = 0$ then $\mathbf{dF} = 0$.
The Faraday is therefore a closed 2-form. Thus,

$$\mathbf{F} = \mathbf{dA} \tag{327}$$

where $\mathbf{A}$ is the 4-potential and $\mathbf{ddA} = 0$. Since $\mathbf{F} = \mathbf{dA}$ we can write,

$$F_{\mu\alpha} = \frac{\partial A_\alpha}{\partial x^\mu} - \frac{\partial A_\mu}{\partial x^\alpha} \tag{328}$$

$$\mathbf{A} = A_0 \mathbf{d}t + A_1 \mathbf{d}x + A_2 \mathbf{d}y + A_3 \mathbf{d}z \tag{329}$$

Definition of Dual of 2-form $\mathbf{F}$,

$$\mathbf{F} = E_x \mathbf{d}x \wedge \mathbf{d}t + E_y \mathbf{d}y \wedge \mathbf{d}t + ... + B_z \mathbf{d}x \wedge \mathbf{d}y \tag{330}$$

Dual,

$$*\mathbf{F} = -B_x \mathbf{d}x \wedge \mathbf{d}t - ... + E_y \mathbf{d}z \wedge \mathbf{d}x + E_z \mathbf{d}x \wedge \mathbf{d}y \tag{331}$$

With this definition we can also show that,

$$* * \mathbf{F} = -\mathbf{F} \tag{332}$$

From Maxwell's equations we have the electrostatic law,

$$\nabla \cdot \mathbf{E} = 4\pi\rho \tag{333}$$

62

and the electrodynamic equation,

$$\frac{\partial \mathbf{E}}{\partial t} - \nabla \times \mathbf{B} = -4\pi \mathbf{J} \qquad (334)$$

Now define the "4-current" J,

$$J^0 = \rho = \text{charge density}$$
$$(J^1, J^2, J^3) = \text{components of current density} \qquad (335)$$

Then Maxwell's electrostatic law and electrodynamic law can be encapsulated in the 2-form expression,

$$\mathbf{d} * \mathbf{F} = 4\pi * J \qquad (336)$$

This can also be expressed as $\nabla \cdot \mathbf{F} = 4\pi * J$.

Thus all of Maxwell's equations can be written using 2-forms in the following two equations:

$$\mathbf{dF} = 0$$
$$\mathbf{d} * \mathbf{F} = 4\pi * J \qquad (337)$$

Chapter 11

Differential Forms and Curvature

11.1 Preliminaries

The following is based on Flanders [4]. We will paraphrase and quote from Flanders in the derivations to follow. Note that the wedge operator "$\wedge$" is implied and due consideration to the sign should be given.

We are familiar in rectangular coordinates with

$$d\mathbf{x} = dx\mathbf{i} + dy\mathbf{j} + dz\mathbf{k} \qquad (338)$$

In orthogonal coordinates in general,

$$d\mathbf{x} = \sigma_1 \mathbf{e_1} + \sigma_2 \mathbf{e_2} + \sigma_3 \mathbf{e_3} \qquad (339)$$

For example in spherical coordinates,

$$dr = d\rho \mathbf{e}_\rho + \rho d\theta \mathbf{e}_\theta + \rho \sin\theta d\phi \mathbf{e}_\phi \tag{340}$$

Hence,

$$\begin{aligned} \sigma_1 &= d\rho d\rho \\ \sigma_2 &= \rho d\theta \\ \sigma_3 &= \rho \sin\theta d\phi \end{aligned} \tag{341}$$

σ is known as a one-form.

Now,

$$d\mathbf{e_i} = \omega_{i1}\mathbf{e_1} + \omega_{i2}\mathbf{e_2} + \omega_{i3}\mathbf{e_3} \tag{342}$$

This is a very important expression. It states that by moving an infinitesimal amount the unit coordinate $\mathbf{e_i}$ changes (magnitude and direction) and its change depends on all unit coordinates weighted by differential factors.

Thus, ω_{ik} is also a 1-form.

Since $\mathbf{e_i} \cdot \mathbf{e_k} = \delta_{ik}$, taking the differential, we have,

$$d\mathbf{e_i} \cdot \mathbf{e_k} + \mathbf{e_i} \cdot d\mathbf{e_k} = 0 \tag{343}$$

Substituting for the $d\mathbf{e_i}$ and $d\mathbf{e_i}$ we get,

$$\omega_{ik} + \omega_{ki} = 0, \omega_{ii} = 0 \tag{344}$$

Introduce matrix notation,

$$\mathbf{e} = \begin{bmatrix} \mathbf{e_1} & \mathbf{e_2} & \mathbf{e_3} \end{bmatrix}^T \tag{345}$$

$$\sigma = \begin{bmatrix} \sigma_1 & \sigma_2 & \sigma_3 \end{bmatrix} \tag{346}$$

$$\mathbf{\Omega} = ||\omega_{ij}|| \tag{347}$$

Where $\mathbf{\Omega}$ is a 3×3 matrix.

From $d(d\mathbf{x}) = 0$ we have,

$$d\sigma \mathbf{e} - \sigma d\mathbf{e} = 0 \tag{348}$$

$$d\sigma \mathbf{e} - \sigma \mathbf{\Omega} \mathbf{e} = 0 \tag{349}$$

$$(d\sigma - \sigma \mathbf{\Omega})\mathbf{e} = 0 \tag{350}$$

Since $\mathbf{e_i}$ are linearly independent,

$$d\sigma = \sigma \mathbf{\Omega} \tag{351}$$

64

Similarly $d(d\mathbf{e}) = 0$. From which,

$$0 = d\mathbf{\Omega}\mathbf{e} - \mathbf{\Omega}d\mathbf{e} = \left(d\mathbf{\Omega} - \mathbf{\Omega}^2\right)\mathbf{e} \tag{352}$$

Or,

$$d\mathbf{\Omega} = \mathbf{\Omega}^2 \tag{353}$$

Note that equation (344) can be written as,

$$\mathbf{\Omega} + \mathbf{\Omega}^T = 0 \tag{354}$$

11.2   Hypersurfaces

Note: This section is a prelude to Riemannian geometry.

From [4]:

Definition A *hypersurface* is an n-dimensional manifold $\mathbf{M}$ embedded in $\mathbf{E}^{n+1}$. Denote a moving point on $\mathbf{M}$ by $\mathbf{x}$. Our study is local so we pick a definite unit normal $\mathbf{n}$ at each point $\mathbf{x}$ of $\mathbf{M}$. The map $\mathbf{x} - > \mathbf{n}$ is a smooth map on $\mathbf{M}$ into S^n.

The tangent space at $\mathbf{x}$ is an n-dimensional Euclidean Space; we pick an orthonormal basis for it $\mathbf{e_1}, \cdots \mathbf{e_n}$. Thus at $\mathbf{x}$, the vectors $\mathbf{e_1}, \cdots \mathbf{e_n}, \mathbf{n}$ make up an orthonormal basis of E^{n+1}. Since $d\mathbf{x}$ is in the tangent space we have,

$$d\mathbf{x} = \sigma_1\mathbf{e_1} + \sigma_2\mathbf{e_2} + \sigma_3\mathbf{e_3} \tag{355}$$

where $\sigma_1, \cdots, \sigma_n$ are one-forms on $\mathbf{M}$.
From the relations,

$$\begin{aligned}
\mathbf{e_i} \cdot \mathbf{e_k} &= \delta_{ik} \\
\mathbf{e_i} \cdot \mathbf{n} &= 0 \\
\mathbf{n} \cdot \mathbf{n} &= 1
\end{aligned} \tag{356}$$

we deduce

$$\begin{aligned}
d\mathbf{e_i} \cdot \mathbf{e_k} + \mathbf{e_i} \cdot d\mathbf{e_k} &= 0 \\
d\mathbf{e_i} \cdot \mathbf{n} + \mathbf{e_i} \cdot d\mathbf{n} &= 0 \\
\mathbf{n} \cdot d\mathbf{n} &= 0
\end{aligned} \tag{357}$$

and so

$$d\mathbf{e_i} = \sum \omega_{ij}\mathbf{e_j} - \omega_i\mathbf{n} \tag{358}$$

$$d\mathbf{n} = \sum \omega_i\mathbf{e_i} \tag{359}$$

where ω_{ij} , ω_i are one-forms on $\mathbf{M}$ and

$$\omega_{ij} + \omega_{ji} = 0 \tag{360}$$

65

Note that $d\mathbf{n}$ lies in the tangent space (since $\mathbf{n}d\mathbf{n} = 0$). Thus,

$$d\mathbf{n} = \sum \omega_i \mathbf{e_i} \tag{361}$$

In matrix form,

$$d\begin{pmatrix} \mathbf{e} \\ \mathbf{n} \end{pmatrix} = \begin{pmatrix} \mathbf{\Omega} & -\omega^T \\ \omega & 0 \end{pmatrix} \begin{pmatrix} \mathbf{e} \\ \mathbf{n} \end{pmatrix} \tag{362}$$

We also have,

$$d\mathbf{x} = \sigma\mathbf{e} \tag{363}$$

$$\mathbf{\Omega} + \mathbf{\Omega}^T = 0 \tag{364}$$

Lets take exterior derivatives. In what follows we will omit "$\wedge$". Following [4]:

$$\begin{aligned} 0 = d(d\mathbf{x}) &= (d\sigma)\mathbf{e} - \sigma(dd\mathbf{e}) \\ &= (d\sigma)\mathbf{e} - \sigma(\mathbf{\Omega}\mathbf{e} - \omega^T\mathbf{n}) \\ &= d\sigma - \sigma\mathbf{\Omega})\mathbf{e} + \sigma^T\omega\mathbf{n} \end{aligned} \tag{365}$$

with

$$d\sigma = \sigma\mathbf{\Omega} \tag{366}$$

$$d\sigma^T\omega = 0 \tag{367}$$

$$\begin{aligned} 0 &= d\left[d\begin{pmatrix} \mathbf{e} \\ \mathbf{n} \end{pmatrix} \right] \\ &= \begin{pmatrix} d\mathbf{\Omega} & -d\omega^{\mathbf{T}} \\ d\omega & 0 \end{pmatrix} \begin{pmatrix} \mathbf{e} \\ \mathbf{n} \end{pmatrix} - \begin{pmatrix} \mathbf{\Omega} & -\omega^T \\ \omega & 0 \end{pmatrix} d\begin{pmatrix} \mathbf{e} \\ \mathbf{n} \end{pmatrix} \\ &= \begin{pmatrix} d\mathbf{\Omega} & -d\omega^{\mathbf{T}} \\ d\omega & 0 \end{pmatrix} \begin{pmatrix} \mathbf{e} \\ \mathbf{n} \end{pmatrix} - \begin{pmatrix} \mathbf{\Omega} & -\omega^T \\ \omega & 0 \end{pmatrix}^2 \begin{pmatrix} \mathbf{e} \\ \mathbf{n} \end{pmatrix} \\ & \quad \begin{pmatrix} d\mathbf{\Omega} - \mathbf{\Omega}^2 + \omega^{\mathbf{T}}\omega & -(d\omega)^T + \mathbf{\Omega}^T\omega \\ d\mathbf{\Omega} - \omega\mathbf{\Omega} & 0 \end{pmatrix} \begin{pmatrix} \mathbf{e} \\ \mathbf{n} \end{pmatrix} \end{aligned} \tag{368}$$

$$d\mathbf{\Omega} - \mathbf{\Omega}^2 + \omega^{\mathbf{T}}\omega = 0 \tag{369}$$

$$d\omega = \omega\mathbf{\Omega} \tag{370}$$

66

We define a skew-symmetric matrix of two-forms,

$$\Theta = ||\theta_{ij}|| = d\mathbf{\Omega} - \mathbf{\Omega}^2 \qquad (371)$$

Summary of results [4]:

$$
\begin{aligned}
d\sigma &= \sigma\mathbf{\Omega} &\qquad (372)\\
\mathbf{\Omega} + \mathbf{\Omega}^T &= 0 &\qquad (373)\\
\sigma^T \omega &= 0 &\qquad (374)\\
d\omega &= \omega\mathbf{\Omega} &\qquad (375)\\
\Theta + \omega^T \omega &= 0 &\qquad (376)
\end{aligned}
$$

Suppose one has a function $\mathbf{v} = \mathbf{v}(y, z, ...)$ where $\mathbf{v}$ is always a tangent vector to $\mathbf{M}$.

"How does an observer constrained to $\mathbf{M}$ observe the motion of $\mathbf{v}$?" [4]

Now,

$$\mathbf{v} = \sum c_i \mathbf{e}_i \qquad (377)$$

c_i are functions.

$$
\begin{aligned}
d\mathbf{v} &= \sum dc_i \mathbf{e}_i + \sum c_i (\sum w_{ij}\mathbf{e}_i - w_i \mathbf{n}) &\qquad (378)\\
&= \sum (dc_i + \sum c_i w_{ij})\mathbf{e}_j - (\sum c_i w_i)\mathbf{n} &\qquad (379)
\end{aligned}
$$

where we have used $d\mathbf{e_i} = \sum \omega_{ij}\mathbf{e_j} - \omega_i \mathbf{n}$ from (358).

From [4]

Our observer who is constrained to move in the hypersurface $\mathbf{M}$ cannot "see" the motion of $\mathbf{v}$ which takes place in the direction normal to $\mathbf{M}$; he sees only the tangential motion of $\mathbf{v}$. Consequently he believes $\mathbf{v}$ is motionless provided,

$$(dc_j + \sum c_i w_{ij})\mathbf{e}_j = 0 \qquad (380)$$

That is ,

$$dc_j + \sum c_i w_{ij} = 0 \; (j = 1, ..., n). \qquad (381)$$

A vector function for which these equations are valid is said to move by **parallel displacement**

The following can be checked. If $\mathbf{v} = \mathbf{v}(y, z, ...)$ and $\mathbf{w} = \mathbf{w}(y, z, ...)$ are two such vector-valued functions which are compatible for each point $(y, z, ...)$ in the parameter space $\mathbf{v}$ and $\mathbf{w}$ are tangent at the same point of $\mathbf{M}$ and each moves by parallel displacement, then $\mathbf{v} \cdot \mathbf{w}$ is constant.

In particular, $|\mathbf{v}|^2 = \mathbf{v} \cdot \mathbf{v}$ is constant.

Let $P = P(s)$ be a curve on $\mathbf{M}$ parameterized by its arc length s so that

$$\mathbf{t} = \mathbf{t}(s) = \frac{dP}{ds} \tag{382}$$

is a unit tangent vector. The curve is called a **geodesic** provided $\mathbf{t}$ moves by parallel displacement.

Recall,

$$d\mathbf{n} = \sum w_i \mathbf{e}_i \tag{383}$$

and

$$d\mathbf{n} \cdot \mathbf{n} = 0 \tag{384}$$

Noting that $d\mathbf{n}$ is in the tangent space. We also have,

$$w_i = \sum b_{ij} \sigma_i \tag{385}$$

We have computed our structure equations:

$$dP = \sigma \mathbf{e} \tag{386}$$
$$d\mathbf{e} = \mathbf{\Omega}\mathbf{e} \tag{387}$$
$$\mathbf{\Omega} + \mathbf{\Omega}^T = 0 \tag{388}$$

We have,

$$d\sigma = \sigma\mathbf{\Omega} \tag{389}$$

Compute,

$$d^2\mathbf{e} = d(d\mathbf{e}) = d(\mathbf{\Omega}\mathbf{e}) = (d\mathbf{\Omega})\mathbf{e} - \mathbf{\Omega}(d\mathbf{e}) \tag{390}$$

$$d^2\mathbf{e} = (d\mathbf{\Omega} - \mathbf{\Omega}^2)\mathbf{e} \tag{391}$$

Curvature Matrix

Set,

$$\mathbf{\Theta} = |\theta_{ij}| = (d\mathbf{\Omega} - \mathbf{\Omega}^2) \tag{392}$$

the **curvature matrix** which appears from the symbolic equation,

$$d^2\mathbf{e} = \mathbf{\Theta}\mathbf{e} \tag{393}$$

as representing a "second derivative" – exactly how one thinks of curvature in elementry differential geometry.

Continue,

$$d\sigma = \sigma \mathbf{\Omega} \tag{394}$$

$$
\begin{aligned}
0 &= d(d\sigma) = (d\sigma)\mathbf{\Omega} - \sigma(d\mathbf{\Omega}) \tag{395}\\
&= (\sigma\mathbf{\Omega})\mathbf{\Omega} - \sigma(d\mathbf{\Omega}) \tag{396}
\end{aligned}
$$

hence,

$$\sigma\mathbf{\Theta} = 0 \tag{397}$$

From,

$$\mathbf{\Theta} = d\mathbf{\Omega} - \mathbf{\Omega}^2 \tag{398}$$

we have

$$
\begin{aligned}
d\mathbf{\Theta} &= d(d\mathbf{\Omega}) - d(\mathbf{\Omega}^2) \tag{399}\\
&= 0 - (d\mathbf{\Omega})\mathbf{\Omega} + \mathbf{\Omega}(d\mathbf{\Omega}) \tag{400}\\
&= -(\mathbf{\Theta} + \mathbf{\Omega}^2)\mathbf{\Omega} + \mathbf{\Omega}(\mathbf{\Theta} + \mathbf{\Omega}^2) \tag{401}
\end{aligned}
$$

hence,

$$d\mathbf{\Theta} = \mathbf{\Omega}\mathbf{\Theta} - \mathbf{\Theta}\mathbf{\Omega} \tag{402}$$

which comprises the Bianchi identity.

11.3 Reimann Curvature Tensor

The θ_{ij} are two-forms which may be written

$$\theta_{ij} = \frac{1}{2}\sum R_{ijkl}\sigma_k\sigma_l \tag{403}$$

which defines the Reimann curvature Tensor. We have,

$$
\begin{aligned}
R_{ijkl} + R_{ijlk} &= 0 \tag{404}\\
R_{ijkl} + R_{jilk} &= 0 \tag{405}
\end{aligned}
$$

The relation $\sigma\mathbf{\Theta} = 0$, or,

$$\sum R_{ijkl}\sigma_j\sigma_k\sigma_l = 0 \tag{406}$$

is equivalent to

$$R_{ijkl} + R_{iklj} + R_{iljk} = 0 \tag{407}$$

69

11.4 Christoffel Symbols

Consider the definition of Γ_{ijk} :

$$\omega_{ij} = \sum \Gamma_{ijk}\sigma_k \tag{408}$$

Note σ_i form a basis. Note, $\boldsymbol{\Omega} = ||w_{ij}||$ and $d\mathbf{e}_i = \sum \omega_{ij}\mathbf{e}_j$ and $d\mathbf{e} = \boldsymbol{\Omega}\mathbf{e}$.
  Now,

$$d\sigma_i = \frac{1}{2}\sum c_{ijk}\sigma_j\sigma_k \tag{409}$$

Note $\sigma_j \wedge \sigma_j = 0$ so no C_{ijj} terms.
  Also

$$c_{ijk} + c_{ikj} = 0 \tag{410}$$

Continuing,

$$d\sigma_i = \frac{1}{2}\sum c_{ijk}\sigma_j\sigma_k = \sum \sigma_j\omega_{ji} \tag{411}$$

$$= \sum \sigma_j \sum \Gamma_{jik}\sigma_k = \frac{1}{2}\sum (\Gamma_{ijk} - \Gamma_{kij})\sigma_j\sigma_k \tag{412}$$

So,

$$\Gamma_{ijk} - \Gamma_{kij} = c_{ijk} \tag{413}$$

we have,

$$c_{ijk} + c_{ikj} = 0 \tag{414}$$
$$\Gamma_{ijk} - \Gamma_{kij} = 0 \tag{415}$$
$$\Gamma_{ijk} - \Gamma_{kij} = c_{ijk} \tag{416}$$

which has solution,

$$\Gamma_{ijk} = \frac{1}{2}(c_{kij} - c_{jki} - c_{ijk}) \tag{417}$$

Compare with the following definition in terms of the metric:

$$\Gamma_{\mu\beta\gamma} = \frac{1}{2}(g_{\mu\beta,\gamma} + g_{\mu\gamma,\beta} - g_{\beta\gamma,\mu}) \tag{418}$$

11.5 Examples

Spherical Coordinate System.

$$\mathbf{x} = r\sin\phi\cos\theta\mathbf{i} + r\sin\phi\sin\theta\mathbf{j} + r\cos\phi\mathbf{k} \tag{419}$$

$$d\mathbf{x} = \frac{\partial\mathbf{x}}{\partial r}dr + \frac{\partial\mathbf{x}}{\partial\theta}d\theta + \frac{\partial\mathbf{x}}{\partial\phi}d\phi \tag{420}$$

$$d\mathbf{x} = (dr)\mathbf{e}_1 + (rd\phi)\mathbf{e}_2 + (r\sin\phi d\theta)\mathbf{e}_3 \tag{421}$$

70

$$\mathbf{e}_1 \quad = \quad \sin\phi\cos\theta\mathbf{i} + \sin\phi\sin\theta\mathbf{j} + \cos\phi\mathbf{k} \tag{422}$$

$$\mathbf{e}_2 \quad = \quad \cos\phi\cos\theta\mathbf{i} + \cos\phi\sin\phi\mathbf{j} - \sin\phi\mathbf{k} \tag{423}$$

$$\mathbf{e}_3 \quad = \quad -\sin\theta\mathbf{i} + \cos\theta\mathbf{j} + 0\mathbf{k} \tag{424}$$

Thus,

$$\sigma_1 \quad = \quad dr \tag{425}$$

$$\sigma_2 \quad = \quad rd\phi \tag{426}$$

$$\sigma_3 \quad = \quad r\sin\phi d\theta \tag{427}$$

Differentiating,

$$d\mathbf{e}_1 \quad = \quad (d\phi)\mathbf{e}_2 + (\sin\phi d\theta)\mathbf{e}_3 \tag{428}$$

$$d\mathbf{e}_2 \quad = \quad (-d\phi)\mathbf{e}_1 + (\cos\phi d\theta)\mathbf{e}_3 \tag{429}$$

$$d\mathbf{e}_3 \quad = \quad (-\sin\phi d\theta)\mathbf{e}_1 + (-\cos\phi d\theta)\mathbf{e}_2 \tag{430}$$

Now ,

$$d\mathbf{e} = \boldsymbol{\Omega}\mathbf{e} \tag{431}$$

$\boldsymbol{\Omega}$ is skew-symmetric.
Finally,

$$\Omega = \begin{pmatrix} 0 & d\phi & \sin\phi d\theta \\ -d\phi & 0 & \cos\phi d\theta \\ -\sin\phi d\theta & -\cos\phi d\theta & 0 \end{pmatrix} \tag{432}$$

Two Sphere

The metric is:

$$ds^2 = a^2 d\phi^2 + a^2\sin^2\phi d\theta^2 \tag{433}$$

The one-forms are:

$$\sigma_1 = \sin\phi d\theta \tag{434}$$

$$\sigma_2 = d\phi \tag{435}$$

To find Ω we compute $d\sigma$

$$d\sigma_1 = \cos\phi d\phi d\theta \tag{436}$$

$$d\sigma_2 = 0 \tag{437}$$

71

Note that $\theta \wedge \theta = 0$.

We have $d\sigma = \sigma\mathbf{\Omega}$. Thus,

$$(d\sigma_1, d\sigma_2) = (\sigma_1, \sigma_2) \begin{pmatrix} 0 & \cos\phi d\theta \\ -\cos\phi d\theta & 0 \end{pmatrix} \tag{438}$$

Finally,

$$\Omega = \begin{pmatrix} 0 & \cos\phi d\theta \\ -\cos\phi d\theta & 0 \end{pmatrix} \tag{439}$$

Curvature Matrix

$$\mathbf{\Theta} = ||\theta_{ij}|| = d\mathbf{\Omega} - \mathbf{\Omega}^2 \tag{440}$$

$\mathbf{\Omega}^2 = 0_{2\times 2}$ since $d\theta \wedge d\theta = 0$.

$$\mathbf{\Theta} = d\mathbf{\Omega} = \begin{pmatrix} 0 & -(\sin\phi)d\phi d\theta \\ (\sin\phi)d\phi d\theta & 0 \end{pmatrix} \tag{441}$$

Now,

$$\theta_{ij} = \frac{1}{2}\sum R_{ijkl}\sigma_k\sigma_l \tag{442}$$

From which,

$$\theta_{12} = \frac{1}{2}(R_{1212}\sigma_1 \wedge \sigma_2 + R_{1221}\sigma_2 \wedge \sigma_1) \tag{443}$$

However, $R_{1212} + R_{1221} = 0$.

So that,

$$\theta_{12} = R_{1212}\sigma_1 \wedge \sigma_2 \tag{444}$$

Since,

$$\theta_{12} = \sin\phi d\phi d\theta = \sigma_1 \wedge \sigma_2 \tag{445}$$

Then,

$$R_{1212} = 1 \tag{446}$$

11.6 Covariant Exterior Derivative and the Bianchi Identity

The next big step is to obtain an expression for the *covariant exterior derivative* of the curvature matrix $\mathbf{\Theta}$. But first some preliminaries. All due to Lovelock and Rund [8]. We agree with [8] that the power is in combining differential forms and tensor analysis.

72

Consider a contravariant 2-form

$$\Pi^j = A^j_{hk} dx^h \wedge dx^k \tag{447}$$

in which the coefficients A^j_{hk} represent the components of a type (1,2) tensor field. Taking the exterior derivative $d\mathbf{\Pi}$,

$$d\mathbf{\Pi} = \frac{\partial A^j_{hk}}{\partial x^l} dx^l \wedge dx^h \wedge dx^k = \frac{1}{3!} \delta^{rst}_{hkl} \frac{\partial A^j_{rs}}{\partial x^t} dx^h \wedge dx^k \wedge dx^l \tag{448}$$

which is not in general tensorial [8].

"In order to construct a tensorial exterior derivative we have to invoke a connection, and accordingly it is now assumed that X_n is endowed with connection coefficients denoted by Γ^j_{hk}. This allows us to introduce the covariant derivative of A^j_{hj}[8]", which is given by,

$$A^j_{hj;t} = \frac{\partial A^j_{rs}}{\partial x^t} + A^m_{rs}\Gamma^j_{mr} - A^j_{ms}\Gamma^m_{rt} - A^j_{rm}\Gamma^m_{st} \tag{449}$$

Define the covariant exterior derivative of the contravariant 2-form by,

$$D\Pi^j = d\Pi^j + \frac{1}{3!} \delta^{rst}_{hkl} A^m_{rs} \Gamma^j_{mt} dx^h \wedge dx^k \wedge dx^l \tag{450}$$

which is a contravariant 3-form.

Now we need to get this into a more illuminating form.
Define,

$$\omega^j_k = \Gamma^j_{kl} dx^l \tag{451}$$

Note that this is the expression of the 1-form $\mathbf{\Omega}$ which recall since the σ_k formed a basis could be represented as $\omega_{ij} = \sum \Gamma_{ijk}\sigma_k$.

We refer to [8] to obtain the following derived from (450) above,

$$D\Pi^j = d\Pi^j + \Pi^h \wedge \omega^j_h = d\Pi^j + \omega^j_h \wedge \Pi^h \tag{452}$$

This is the covariant exterior derivative of the 1-form Π^j.

Now consider the 2-form Π^j_l,

$$\Pi^j_l = A^j_{lhk} dx^h \wedge dx^k \tag{453}$$

where A^j_{lhk} is a type (1,3) tensor field.

In this case the covariant exterior derivative is (see[8]),

$$D\Pi^j_i = d\Pi^j_l + \omega^j_h \wedge \Pi^h_l - \omega^h_l \wedge \Pi^j_h \tag{454}$$

73

Now for the cool part. Recall that for the curvature matrix Θ

$$\theta_{ij} = \frac{1}{2} \sum R_{ijkl} \sigma_k \sigma_l \tag{455}$$

Take the covariant exterior derivative of the curvature matrix Θ using (454):

$$D\Theta = d\Theta + (\Omega\Theta - \Omega\Theta) \tag{456}$$

However, based on (402) above which is from [4],

$$d\Theta = \Omega\Theta - \Theta\Omega \tag{457}$$

Substituting for $d\Theta$ we obtain,

$$D\Theta = \Omega\Theta - \Theta\Omega - (\Omega\Theta - \Theta\Omega) = 0 \tag{458}$$

$$D\Theta = 0 \tag{459}$$

As [8] points out this is a remarkable identity.

We will show that it is key to wiring geometry to the energy momentum tensor as we shall see later.

Absolute Differential

Consider,

$$D\Pi^j = d\Pi^j + d\Pi^j + \Pi^h \wedge \omega_h^j = d\Pi^j + \omega_h^j \wedge \Pi^h \tag{460}$$

From [8]

> If one regards the components of X^j of a type (1,0) tensor field as representing n 0-forms, one may write
>
> $$DX^j = dX^j + \omega_h^j X^h \tag{461}$$
>
> which by virtue of $\omega_h^j = \Gamma_{hl}^j dx^l$ coincides with the absolute differential of X^j.

Bianchi Identity: Another View

From [7]:

> If we choose a frame which is geodesic at a point x^i, $\partial g_{ij}/\partial x^k$ will vanish at the point and the two Christoffel symbols will therefore also vanish at the point. Thus, all the components of the affinity will vanish at the point and covariant derivatives will reduce to partial derivatives there. It then follows that the partial derivatives $\partial g_{ij}/\partial x^k$ all vanish at the point also (from $g_{;m}^{ir} = 0$).

74

In such a geodesic frame, therefore,

$$R^i_{jkl;m} = \frac{\partial}{\partial x^m}\left(\Gamma^i_{rk}\Gamma^i_{jl} - \Gamma^r_{rl}\Gamma^r_{jk} + \frac{\partial\Gamma^i_{jl}}{\partial x^k} - \frac{\partial\Gamma^i_{jk}}{\partial x^l}\right) \qquad (462)$$

$$= \frac{\partial^2\Gamma^i_{jl}}{\partial x^m\partial x^k} - \frac{\partial^2\Gamma^i_{jk}}{\partial x^m\partial x^l} \qquad (463)$$

In the above we have $\Gamma^i_{jk} = 0$ since we have transformed to a frame such that this is true (Geodesic Frame). Note that this does not mean that the derivative is zero.

Cyclically permuting the indices k,l,m in 462 the following are obtained:

$$R^i_{jkl;m} = \frac{\partial^2\Gamma^i_{jl}}{\partial x^m\partial x^k} - \frac{\partial^2\Gamma^i_{jk}}{\partial x^m\partial x^l} \qquad (464)$$

$$R^i_{jlm;k} = \frac{\partial^2\Gamma^i_{jm}}{\partial x^k\partial x^l} - \frac{\partial^2\Gamma^i_{jl}}{\partial x^k\partial x^m} \qquad (465)$$

$$R^i_{jmk;l} = \frac{\partial^2\Gamma^i_{jk}}{\partial x^l\partial x^m} - \frac{\partial^2\Gamma^i_{jm}}{\partial x^l\partial x^k} \qquad (466)$$

Addition of the three equations yields the following identity:

$$R^i_{jkl;m} + R^i_{jlm;k} + R^i_{jmk;l} = 0 \qquad (467)$$

From [7]:

> But this is a tensor equation and, having been proved true in the geodesic frame, must be true in all frames. Also, since the chosen point can be any point of $\mathbf{R}^N$, it is valid at all points of the space. It is the *Bianchi identity*.

Chapter 12

How Mass-Energy Generates Curvature

The following is from [3] Chapter 17 and ties it all together. We will quote liberally.

Mass is the source of gravity. The density of mass-energy as measured by any observer with 4-velocity **u** is

$$\rho = \mathbf{u} \cdot \mathbf{T} \cdot \mathbf{u} = u^{\alpha} T_{\alpha\beta} u^{\beta} \tag{468}$$

Therefore the stress-energy tensor **T** is the frame-independent "geometric object" that must act as the source of gravity.

This source, this geometric object, is not an arbitrary symmetric tensor. It must have zero divergence

$$\nabla \cdot \mathbf{T} = 0, \tag{469}$$

because only so can the law of conservation of momentum-energy be upheld.

Place this source, **T**, on the righthand side of the equation for the generation of gravity. On the left hand side will stand a geometric object that characterizes gravity. That object, like T, must be a symmetric, divergence-free tensor; and if it is to characterize gravity, it must be built out of the geometry of spacetime and nothing but that geometry. Give this object the name, "Einstein tensor" and denote it by **G**, so that the equation for the generation of gravity reads

$$\mathbf{G} = \kappa \mathbf{T}. \tag{470}$$

The vanishing of the divergence $\nabla \cdot \mathbf{G}$ is not to be regarded as a consequence of $\nabla \cdot \mathbf{T} = 0$. Rather, the obedience of all matter and fields to the conservation law $\nabla \cdot \mathbf{T} = 0$ is to be regarded (1) as a consequence of the way they are wired into the geometry of spacetime, and therefore (2) as required and enforced by an *automatic* conservation law , or *identity*, that holds for any smooth Riemannian spacetime whatsoever, physical or not: $\nabla \cdot \mathbf{G} == 0$. Accordingly, look for a symmetric tensor G that is an "automatically conserved measure of the curvature of spacetime" in the following sense:

(1) **G** vanishes when spacetime is flat.

(2) **G** is contructed from the Riemann curvature tensor and the metric, and from nothing else.

(3) **G** is distinguished from other tensors which can be built from **Riemmann** and **g** by the demands (i) that it be linear in **Riemmann**, as befits any natural measure of curvature; (ii) that, like T, it be symmetric and of second rank; and (iii) that it have an automatically vanishing divergence,

$$\nabla \cdot \mathbf{G} == 0. \tag{471}$$

Apart from a multiplicative constant, there is only one tensor that satisfies these requirements of being an automatically conserved, second-rank tensor, linear in the curvature, and of vanishing when spacetime is flat. It is the Einstein curvature tensor, G, expresses in terms of the Ricci curvature tensor:

$$R_{\mu\nu} = R^{\alpha}_{\mu\alpha\nu}, \tag{472}$$

$$G_{\mu\nu} = R_{\mu\nu} - \frac{1}{2}g_{\mu\nu}R. \tag{473}$$

Recall,

$$R^{\beta}_{\delta} == R^{\mu\beta}_{\mu\delta} \tag{474}$$

$$R = R^{\beta}_{\beta} \tag{475}$$

where R is referred to as the *Curvature Scalar*.

The vanishing of $\nabla \cdot \mathbf{G}$ follows as a consequence of the elementary principle of topology that "the boundary of a boundary is zero".

We might add that in addition to the quote above from [3] that "the boundary of a boundary is zero", in this case, it is a consequence of (459):

$$D\mathbf{\Theta} = 0 \tag{476}$$

Chapter 13

Final Comments

The relationship (470) can be written as:

$$G^{\sigma\tau} = 8\pi T^{\sigma\tau} \tag{477}$$

where we have replaced κ with 8π. It is important to note that (477) reduces to the Newtonian limit specifically when "gravity is weak, the relative motion of the sources is much slower than the speed of light c, and the material stresses are much smaller than the mass-energy density (in units where $c = 1$)."[12].

For the derivation of the Newtonian limit see [12]. For applications of (477)to weak gravity conditions, the Scharzschild solution, Cosmology and Black Holes see [12], [3], and [2] among others.

Bibliography

[1] G.K. Batchelor. *An Introduction To Fluid Dynamics*. Cambridge Mathematics Library, 1967,1973,2000.

[2] Sean Carroll. *SpaceTime and Geometry*. Pearson Addison Wesley, 2004.

[3] John Archibald Wheeler Charles W. Misner, Kip S. Thorne. *Gravitation*. W.H. Freeman and Company, 1973.

[4] Harley Flanders. In *Differential Forms with Applications to the Physical Sciences*. Dover Publications, 1963,1989.

[5] Charles Fox. Late professor of mathematics. In *An Introduction to the Calculus of Variations*. Dover Edition, 1950, 1963.

[6]   David C Kay. *Tensor Calculus*. McGraw-Hill, 1988.

[7]   D. F. Lawden. In *Introduction to Tensor Calculus, Relativity and Cosmology*. Dover Publications, 1982,2002.

[8]   David Lovelock and Hanno Rund. *Tensors, Differential Forms, and Variational Principles*. Dover Publications, 1989,1975.

[9]   Peter Russer. In *Electromagnetics, Microwave Circuit and Antenna Design for Communications Engineering, Second Edition*. Artech House, 2006.

[10] H. M. Schey. *Div Grad Curl and all that*. W.W.Norton and Company, third edition, 1997.

[11] Peter Szekeres. *Modern Mathematical Physics*. Cambridge University Press, 2004.

[12] Robert M. Wald. *General Relativity*. University of Chicago Press, 1984.

[13] Robert C Wrede. *Introduction to Vector and Tensor Analysis*. Dover Publications, 1963,1972.

[14] Daniel Zwillinger. In *Standard Mathematical Tables and Formulae, 31st Edition*. CRC Press, 2003.